Archaeology of East Asian Shipbuilding

UNIVERSITY PRESS OF FLORIDA

Florida A&M University, Tallahassee
Florida Atlantic University, Boca Raton
Florida Gulf Coast University, Ft. Myers
Florida International University, Miami
Florida State University, Tallahassee
New College of Florida, Sarasota
University of Central Florida, Orlando
University of Florida, Gainesville
University of North Florida, Jacksonville
University of South Florida, Tampa
University of West Florida, Pensacola

Archaeology of East Asian Shipbuilding

JUN KIMURA

University Press of Florida
Gainesville · Tallahassee · Tampa · Boca Raton
Pensacola · Orlando · Miami · Jacksonville · Ft. Myers · Sarasota

Printed in the United States of America on acid-free paper

This book may be available in an electronic edition.

21 20 19 18 17 16 6 5 4 3 2 1

Library of Congress Cataloging-in-Publication Data
Names: Kimura, Jun (Marine archaeologist), author.
Title: Archaeology of East Asian shipbuilding / Jun Kimura.
Description: Gainesville : University Press of Florida, [2016] | 2016. |
Includes bibliographical references and index.
Identifiers: LCCN 2015041909 | ISBN 9780813061184 (alk. paper)
Subjects: LCSH: Shipbuilding—East Asia—History. |
Shipbuilding—China—History. | Shipbuilding—Korea—History. |
Shipbuilding—Japan—History. | Underwater archaeology—East Asia. | East
Asia—Antiquities.
Classification: LCC VM99 .K56 2016 | DDC 623.820095/09009—dc23
LC record available at http://lccn.loc.gov/2015041909

The University Press of Florida is the scholarly publishing agency for the State University System of Florida, comprising Florida A&M University, Florida Atlantic University, Florida Gulf Coast University, Florida International University, Florida State University, New College of Florida, University of Central Florida, University of Florida, University of North Florida, University of South Florida, and University of West Florida.

University Press of Florida
15 Northwest 15th Street
Gainesville, FL 32611-2079
http://www.upf.com

Contents

Illustrations

Figures

Maps

Tables

Preface

Maritime archaeology examines past human interaction with seas, rivers, and fresh waters. It provides the means to assess various archaeological remains, such as ship cargo and hull remains, within maritime archaeology. However, apart from ship cargo studies, the East Asian regions demonstrate a lack of research on ship hull remains. My broad aim in this book is therefore to provide a better understanding of the history of ship structures and construction methods that formed shipbuilding traditions within the East Asian region over the centuries.

Two types of data are invoked as resources. The first consists of historical data related to the time period of the ship remains addressed. Written information and pictorial resources describing ship configurations, characteristics, and construction techniques have been reviewed. The contents include historical texts such as travel records, written histories of the period, and iconographic paintings and historical accounts of environmental conditions. Most of these have previously been studied by Chinese and Japanese researchers as well as others working outside East Asia. They have reviewed the history of East Asian ships and shipbuilding. Yet the outcomes emerging from the respective different groups have not been sufficiently addressed together. Hence, while the historical data in this book are complementary to the archaeological approach, part of my aim is also to integrate the historical material collectively into archaeological study of East Asian shipbuilding traditions.

The primary resource for addressing the issue is archaeological data. This book contains the most informative compilation of data on excavated ships ever produced for East Asia. An inclusive data set of the ship remains in East Asia is presented based on the Shipwreck ASIA project supported by the Toyota Foundation (Kimura 2010b). This project has pursued nautical archaeology highlighting regional themes and issues and is designed to collect accessible data on excavated ships in East Asia. I have collected archaeological data including unpublished photographs and drawings at repositories in the Western Australian Museum, Quanzhou Maritime Museum (Museum of Overseas Communication History) in China, National Research Institute of Maritime Cultural Heritage (National Maritime Museum) in Korea, and Buried Cultural Property Investigation Center of Takashima in Japan. Of these institutions, the three in East Asia are all in the position of custodians for the remains of ships that were constructed in the East China Sea region. Archaeological work conducted on these ship remains has consisted of recording ship lines; detailed recording of hull features and components, including taking measurements and identifying fastener patterns, joinery patterns, and construction sequence; sampling of timbers for wood species analysis and radiocarbon dating; X-ray and CT scan analyses of the sampled wooden timbers; and metallurgical studies to assess the quality of iron remains.

The data collected allowed me to trace hull structure and construction methods in detail over time, which is important for clarifying technological evolution in East Asian shipbuilding traditions. The focus on the hull component is pursued with a theory highlighting technological innovations and diffusions, rather than a focus on a linear evolutionary development model. The concept of "hybridization" in shipbuilding technology, initially highlighted by Pierre-Yves Manguin (1984) with an insistence on the "South China Sea shipbuilding tradition," is the basis for my perspective. In this book, however, I aim to provide a more holistic view by linking more broadly the shipbuilding technologies developed in the Yellow Sea, East China Sea, and South China Sea and explaining the historical dynamics of the shipbuilding traditions in East Asia over the centuries. This archaeological research thus pursues the distinctive regional shipbuilding traditions yet also adds a new perspective to the existing body of theoretical thought on the historical development of shipbuilding traditions in maritime archaeology.

Acknowledgments

This book is based on my research in the Maritime Archaeology Program at Flinders University. That work would never have been completed without the support of the organizations and individuals recognized here. It was fully supported by the Endeavour Postgraduate Award provided by the Australian government from 2007 to 2011, with administrative assistance from Austraining International. The Mishima Kaiun Memorial Foundation provided a research grant to conduct fieldwork overseas. Additional funding came from numerous research grants available through Flinders University, and as mentioned in the preface, some of the data derive from the outcomes of a project financially assisted by the Toyota Foundation.

While doing fieldwork in China, I obtained full support from the staff of the Quanzhou Maritime Museum (Museum of Overseas Communication History), and I would like to express my gratitude to Dr. Yuling Ding and Feng Wang. During my fieldwork in Korea, the staff of the National Research Institute of Maritime Cultural Heritage (National Maritime Museum) provided in-kind support, and I want to thank Nack-Jun Seong, Chul-han Lee, and Dr. Whan-suk Moon. I greatly benefited from the support of the Matsuura City staff during my research in Takashima. Some of the data presented in this volume were obtained through cooperation with Drs. Shuichi Noshiro and Hisashi Abe of the Forestry and Forest Products Research Institute and Masami Osawa of Kyushu Techno Research Inc. My appreciation extends to experts and members of the following organizations for their insightful advice and generous support: Asian Research Institute of Underwater Archaeology, Australasian Institute for Maritime Archaeology, École Française d'Extrême-Orient, the Department of Maritime Archaeology of the Western Australian Museum, the Institute of Nautical Archaeology at Texas A&M University, the

Field Museum, the Ministry of Marine Affairs and Fisheries of the Republic of Indonesia, National Museum of the Philippines, the Shipbuilding History Research Centre at Wuhan University, and the Thai Underwater Archaeology Division of the Fine Arts Department of Thailand.

My gratitude goes to many people not listed here who helped me during the past decade or more, as some aspects of the book's contents are from excavations in which I participated more than ten years ago. I especially thank Randall Sasaki, Amer Khan, and Nicholas Burningham for their important support during the fieldwork and for providing recommendations and feedback. I owe thanks to colleagues during my PhD candidature and postdoctoral fellowship—Dr. James Hunter, Dr. Debra Shefi, Dr. Jennifer Rodrigues, Ross Anderson, Nicolas Bigourdan, Charlotte Minh Ha Pham, Dr. Michael McCarthy, Dr. Jason Raupp, and Dr. Joseph Christensen. I owe a very important debt to Kelvin Rodrigues. Dr. James Warren of Murdoch University encouraged me to publish my research, and this was directed by Dr. James Delgado and Dr. Hans Van Tilburge of the National Oceanic and Atmospheric Administration. Dr. Mark Staniforth, Dr. Jennifer McKinnon, and Jeremy Green are acknowledged for their supervision and critiques based on their expertise. My heartfelt thanks go to them all.

Dynastic Timeline

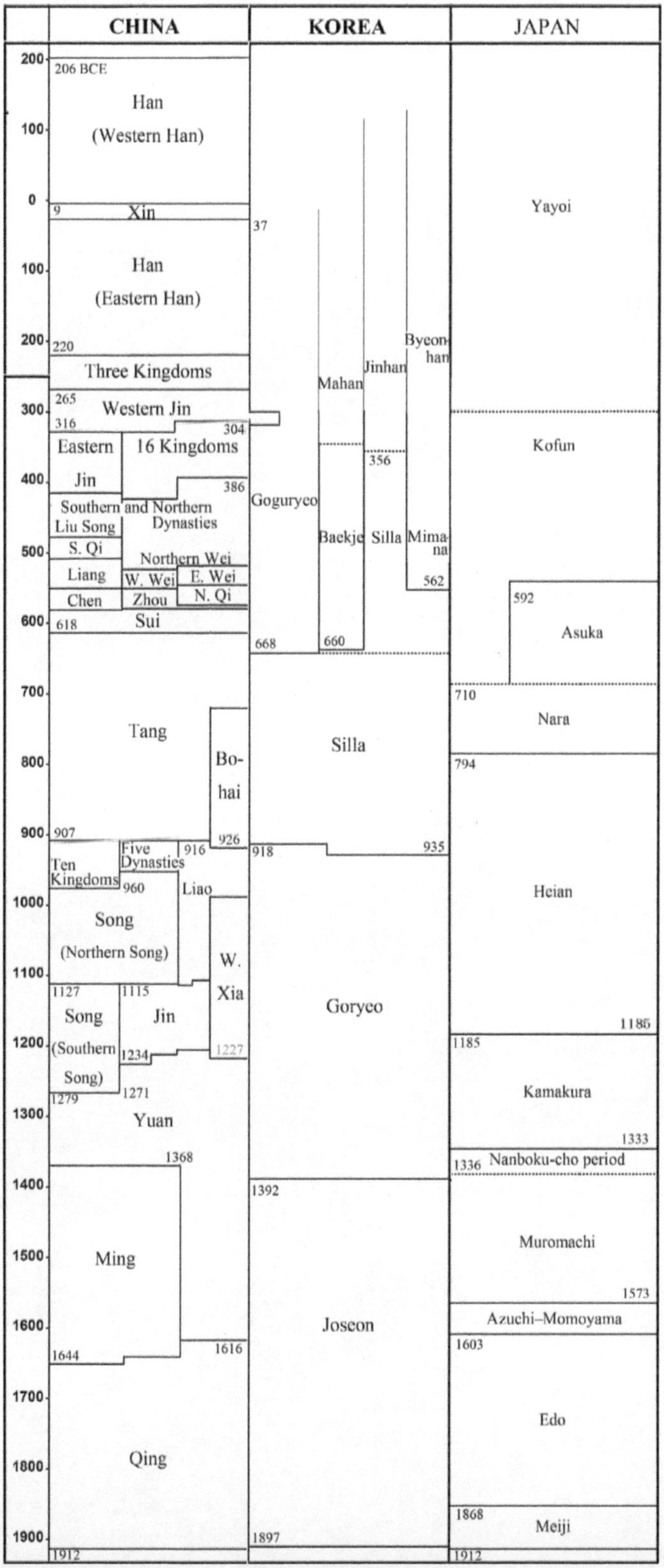

1

Dynamics of East Asian Shipbuilding Traditions

Excavated ships are valuable resources for our understanding of the seafaring past, early human life and behavior on board, and traditional shipbuilding, as well as providing insight into the development of society, technology, ideology, and material culture (Maarleveld 1995; Adams 2001; Gibbins and Adams 2001; Staniforth 2002). The field of archaeology with a specific focus on hull remains has lagged behind in East Asia, compared to the growing body of studies in Europe and the United States in recent decades (Basch 1972; Greenhill 1976; McGrail 1985; Steffy 1994; McGrail 2001). Few attempts have been made by local experts to use information on excavated ships in the region as a data set for thematic research. The chronological framework regarding the development of hull construction techniques and structure is insufficiently understood in regional maritime history. This book is primarily aimed at filling that gap by qualitatively presenting the suite of available data on the excavated hulls of East Asian ships. A further objective is to achieve a more comprehensive understanding of the structural characteristics of ships that once sailed along the coasts of China, Korea, and Japan and those vessels that crossed the waters between these nations. The archaeological data are derived from excavated ships in the regions embraced by the Yellow Sea, the East China Sea, and the South China Sea (map 1.1), ranging from the eleventh to the fifteenth century. Using these data, I explore the long-term dynamics of East Asian shipbuilding traditions with a focus on technological innovations that are relevant to the individual seas.

Historiography of the East Asian Shipbuilding Traditions

Historical study of traditional shipbuilding in East Asia has been contributed by researchers from different fields (e.g., Wade 2003; Van Tilburg

Map 1.1. Locations of ship remains dating to the late thirteenth and early fourteenth centuries and key regions and cities discussed in this volume. (Produced by author, after CraftMAP)

1994, 2007). This allows for a multidisciplinary approach to the theme, referring to historical texts, iconographic resources, and ethnographic material as well as archaeological evidence. Nautical study in East Asia was recognized as an academic discipline by researchers from Europe and the United States having a perception of the shipbuilding tradition outside the Eurocentric tradition. Their interests mainly lay with indigenous boats and ships in China, from primitive watercraft to seagoing ships they termed junks, as well as the chronological development of these vessels (Donnelly 1923, 1924, 1925, 1926, 1930, 1936; Waters 1938, 1939, 1947). George R. G. Worcester (1890–1969), a British officer in the Chinese Maritime Customs Service in the early to mid-twentieth century, made a remarkable record of wooden sailing ships in China, influencing later studies (Sigaut 1960; Needham 1971; Sokoloff 1982). His boat ethnological study provides insight on variations in construction methods for Chinese ships, including the order of construction (Worcester 1941, 1966). His monograph, which records more than 250 watercraft on the Yangtze River, is comprehensive in terms of a detailed study of propulsion, with classification of rigs and sails, different types of rudders and anchors, fastening methods, caulking techniques, cargo capacity, and even sailors' traditional customs and beliefs (Worcester 1971).

Compared to the study of watercraft in China, there are not many studies by non-Asian researchers on the same topics in other East Asian countries. Joseph Needham's work provides descriptive information from a historical document on an early twelfth-century voyage of a Chinese envoy to the Goryeo dynasty of Korea (AD 918–1392), based on the *Xuanhe Fengshi Gaoli Tujing* (Illustrated record of an envoy to Goryeo in the Xuanhe reign-period, 宣和奉使高麗圖經). It deals with the structure of Korean ships in comparison to Chinese ships dating to the same period (Needham 1971). Horace H. Underwood, who lived in Korea as a missionary, conducted an extensive study of boat ethnology of traditional Korean ships (Underwood 1979). His work covers the construction method for local ships, the shipbuilding and transportation industries, and ship carpenters' customs.

More archaeological findings about coastal traders and seagoing ships in East Asia have emerged since the 1970s. Ship remains in China have been extensively reviewed as part of an inclusive study of archaeological remains of watercraft in the world (Steffy 1994; McGrail 2001). This resulted from the locating of ship remains and shipwrecks in East Asia during the last quarter of the twentieth century. Beginning in the 1970s, researchers from the Western Australian Museum directly assessed the sites and introduced underwater and nautical archaeological approaches based on Western perspectives into the region (Green 1983a, 1986; Kenderdine 1995; Burningham and Green 1997; Kenderdine 1997; Green et al. 1998; Green 2006).

In the 1980s, the site investigations began to be led by national institutes in China and Korea. The Chinese government established the Underwater

Archaeological Research Center (UARC) under the National Museum of Chinese History (currently the National Museum of China) in 1986, followed by the enactment of the Regulation on the Protection and Administration of Underwater Cultural Relics in 1989 (Zhao 1992; Hoagland 1997, 1999; Zhang 2011; Adams 2013). Numerous surveys have been carried out by UARC in cooperation with overseas organizations, such as the Western Australian Museum and the Institute of Underwater Archaeology in Japan (Atkinson 1990; Green and Clark 1993; Green et al. 1993; Kenderdine 1995). The work of UARC and the newly established National Conservation Center for Underwater Cultural Heritage, however, focuses mainly on the cargoes of shipwrecks, rather than hull structures, even though more than seventy wreck sites have been found and some of the ships, including the Nanhai shipwreck no. 1 and Huaguang Reef shipwreck no. 1, are remarkably well preserved (Zhang 2006a; Shan 2011). A paucity of research on maritime cultural heritage—such as ports, shipwrecks, coastal settlements, and ethnological resources—has been pointed out (Wu 2011). Under the circumstances, over the past few decades, Chinese shipbuilding traditions have been reviewed via excavated ships at terrestrial sites, not by maritime archaeologists but by maritime historians, naval engineers and architects, and a few terrestrial archaeologists (Xi 1989; Wang 2000; Xi 2000; Cultural Relics and Archaeological Institute of Shandong Province et al. 2006).

In Korea, interest in the native watercraft has risen in relatively recent times (Kim 1994; Choi 2006). Remarkable progress has occurred in exploring shipwrecks since the late 1970s, following the discovery of a thirteenth-century Chinese trader known as the Shinan ship. A conservation center at Mokpo city, initially established for conservation of the Shinan shipwreck and later developed as the National Maritime Museum, has been a focal point for underwater archaeological research in Korea. Its researchers have been active in locating previously unknown shipwrecks in the waters off the west coast of the Korean Peninsula. By 2011 they had identified nine, ranging in date from the end of the eleventh century to the fourteenth century (National Maritime Museum of Korea 2005a; Kim 2006; National Research Institute of Maritime Cultural Heritage 2010, 2011b). These ships were local coastal traders of the Goryeo dynasty and have a flat bottom structure, exhibiting similarity in construction.

I examine their structural characteristics in detail later in this book to address the historical interactions between shipbuilding traditions in the Korean Peninsula and Japanese shipbuilding, which has to date been considered a distinctive technology. Researchers in Japan have agreed that the long-term use of dugout canoes was the basis for the emergence of ships endemic to Japan; the types of dugout canoes that developed can even be linked to the traditional Japanese coastal traders commonly used between the seventeenth and nineteenth centuries (Ishii 1957; Shimizu 1968, 1975; Okita 2000b). I reexamine this perspective in light of what has been identified in the archaeological remains in the Korean Peninsula.

It is worth referring to Japanese maritime historians' achievements involving not only ships of Japanese origin but also ships from overseas visiting the country. The topics of previous studies are diverse, but most highlight the heyday of maritime activities by Chinese traders (Oba 1980; Tanaka and Matsuura 1986; Oba 2001; Matsuura 2003; Oba 2003; Matsuura 2008a, 2008b). Among the iconographic sources introduced in these studies, perhaps the most important is the early eighteenth-century picture scroll *Tosen no Zu* (Drawings of Chinese ships, 唐船の図), depicting Chinese traders at the time (Oba 1974, 1991). This scroll complements the archaeological data.

Archaeologically, investigation of underwater sites under the auspices of Japanese government has been less extensive than that of the other two East Asian nations (Hayashida 2006). Indeed, shipwreck archaeology has yet to be recognized in Japan as a subdiscipline of archaeology, similar to most of the other East and Southeast Asian countries (Kimura 2009, 2015). Yet information derived from archaeological work at sites in Japan has proved critical to this book. A representative case is a naval battlefield site related to Kublai Khan's thirteenth-century attempted invasion, which was countered by fleets from the Chinese mainland and the Korean Peninsula. The remnants of the disarticulated ship timbers from the site make up an assemblage of clues for addressing shipbuilding around the site, as I demonstrate later.

Elements and Definition of Seagoing Ships

Of the various types of ships, the historical ships capable of venturing out to sea are of most interest to me. A structural characteristic such as hull shape and the presence of distinctive transverse structures is specifically used here for the definition of such ships and serves as a parameter to orient the discussion of the development of the shipbuilding tradition.

Coastal ships from the Korean Peninsula with a flat bottom arose early in East Asian maritime history. Later, China started to serve the region with various sailing ships designed for long-distance voyages, and their bulkhead structure has been credited as a significant and remarkable Chinese innovation (Needham 1971: 389–92; Basch 1972; Keith and Buys 1981; Green 1998; McGrail 2001: 347–84). Focusing on the bulkhead structure is critical to understanding the seagoing ships that played a major role in maritime activities in East Asian waters.

Development of the bulkhead structure is arguably associated with the evolution of different functions, including partitioning, creating watertight components, and enhancing transverse strength. The installation of partitions or walls inside the hull is common not only in Chinese ships but also in watercraft in other parts of the world throughout the ancient and modern eras (Burningham 1989; Steffy 1994: 52). Small semicircular partitions in the bottom of ancient dugout canoes have been identified as the ancestral form, as evidenced in parts of Eurasia, including Russia, Vietnam, and Japan (Okorokov 1995; Ishii 1995a; Bellwood et al. 2007; Caira 2008). Yet whether the partition boards of these primitive dugout canoes would come under the modern criteria of bulkheads (as vertical components separating spaces in the hull bottom from one another) is disputable. In a dugout canoe, dividing the interior space of the hull is less necessary, which suggests that this kind of semicircular board in primitive watercraft can be distinguished from the underlying idea of a bulkhead used for partition purposes. One can, however, recognize the usefulness of having partition boards to secure private property and space inside later ships, such as those of thirteenth-century China (Needham 1971: 470).

Chinese researchers have proposed, based on historical accounts, that watertight bulkheads began to be used in China as early as the fifth century (Xi et al. 2004; Cai et al. 2010). The introduction of watertight bulkheads in China earlier than in Europe has been considered one of numerous important innovations in naval architecture (Bernard 1844: 8; Young 1867: 31; Yule 1993: 250). However, the time period and the detailed background of the introduction of watertight bulkhead structures and construction methods must be carefully examined with further concrete evidence.

In terms of both function and East Asian shipbuilding traditions, the use of bulkheads as transverse structures for the hull was an important technological innovation in the construction of truly seaworthy traders. Chinese researchers have agreed that bulkhead construction became widespread at least by the Tang dynasty (AD 618–907), given that a few riverine boats dating to that period demonstrate the presence of bulkheads (Xi et al. 2004). Those riverine ships could have been seaworthy enough to sail on the shallow Yellow Sea. The appearance of bulkheads in combination with a keel and hull planking in ships built in the East China Sea region occurred during a later period as a well-established technology for seagoing purposes. Adoption of this shipbuilding technology is evidenced in the remains of seagoing vessels found in China, Korea, and Japan. In later chapters I describe the details of hull remains of seagoing traders, including the Quanzhou ship (dated to the second half of the twelfth century) found in China and the Shinan shipwreck (dated to the first quarter of the fourteenth century) in Korea. These ships are representative cases for detailed structural study, including a fuller assessment of how bulkheads were used in the seagoing ships plying the East China Sea at the time. Disarticulated ship timbers from the Takashima underwater site, where Kublai Khan's fleets were lost, provide comparable data for the examination of shipbuilding technology at a regional perspective. The dominance of bulkhead construction became apparent in the fifteenth century, as seen in the common technology prevalent in ships in the South China Sea from that period (Manguin 1984), such as the seagoing traders excavated in Southeast Asia. My focus on structural elements allows discussion of the broad linkage of the shipbuilding traditions of East Asia and Southeast Asia.

Hybridity in Shipbuilding Traditions

The question arises as to whether a theory in shipwreck archaeology focused on the sequence of construction is fully applicable for East Asia. The existent theory has been widely accepted and is one of the normative approaches for interpreting the development of shipbuilding (Steffy 1994; Hocker 2004). Yet the validity of the approach to shipbuilding traditions outside the Western norm is controversial (Basch 1972; Reinders and Paul 1991). In the current situation in which a relevant theoretical approach to understanding ship construction practices in East Asian traditions remains relatively undeveloped, I pursue a theoretical framework through detailed descriptions of the structural characteristics of multiple ship remains dating back over several centuries as well as an interpretative analysis that locates technological innovations in shipbuilding traditions. The remnants of seagoing ships—including longitudinal and transverse structures, planking, and fastenings—are examined in chronological order, thereby illustrating representative structural features and construction methods of East Asian seagoing vessels in different periods. The technological innovations through history are addressed by focusing on the functional development of these individual components and their relationships to one other. The possibility of intra- and interregional technological integration is identified through the concept of hybridization.

Jeremy Green (1990) and Pierre-Yves Manguin (1993) initially suggested the importance of understanding hull components in detail to explain distinctive shipbuilding technologies used in the adjacent regions of East Asia and Southeast Asia. Examining the regional characteristics of hull components is significant for clarifying the technological variations in ship construction. A bias exists in the discussion of traditional shipbuilding in East Asia, whereby Japan and the Korean Peninsula are likely to be assessed in the context of the cultural sphere of the Chinese civilization instead of independently. Because of this, detailed study of ship structures existing in these regions is often overlooked. The shipbuilding traditions of Japan and Korea are quite distinct from those of China; in particular, bulkheads were not used, but an alternative transverse beam system was developed (Green 1986). This transverse beam system needs to be better identified and related to broader East Asian shipbuilding practices. Through a detailed interpretation of multiple structural components of ship remains, I aim to provide a holistic view of the development of regional shipbuilding technologies in East Asia.

The spread of innovations has traditionally been argued in diffusion theories in archaeology relevant to the study of cultural diffusion models (Johnson 1999). However, the most fundamental consideration for our purposes is the consequence of the dissemination of technology that resulted from the integration of hull construction methods. The South China Sea is a good example for addressing the issue of the resulting hybridization, although certainly it is not the only case of hybridity in historical shipbuilding. The geographical position of the South China Sea contributed to its being a focal area accessed by different groups: during the early period by people from the Indian Ocean regions, and then actively by East Asian seafarers, which continued up to the European colonial period. Cultural hybridity has been considered important in regional study (Reid 2010: 307), and the idea has already been invoked in the study of Southeast Asian shipbuilding (Manguin 1984, 1993, 2010). I pursue the matter more extensively than in previous studies, drawing from additional evidence.

One achievement exemplifying technological innovations from either endogenous or exogenous growth is the hybrid hull, though the innovations used for watercraft may consist of clustering technologies, rather than solely of dispersing technology. Conceptually, this can range from an innovative idea for fashioning a primitive watercraft to adoption of a shipwright's tool that can realize the idea in a better way. Such a technological cluster comprising both tangible and intangible components may not appear to be an assembly of multiple innovations in the primitive types of watercraft but may be more readily visible in the adopting of multiple technologies in hull construction. The technological cluster, in becoming apparent, evolves through integration of hull components. The phenomenon upon reaching this complex stage can be described as hybridization because the ships begin to show hybridity in their configuration.

In some regions of Asia, the appearance of hybrid ships can be observed in iconography and literature. European people arrived in Southeast Asia by the sixteenth century and at some stage recognized ships as having features from different regions. There is a historical drawing of a ship that has been interpreted as a Chinese ship involved in Southeast Asian waters before 1613. However, a later analysis corrected the erroneous identification, revealing this ship to be not Chinese in origin but a ship of local construction in Southeast Asia, known by the Portuguese as a Malay or Javanese "jong (or junco)" (Manguin, quoted in Guy 1980: 17). The depiction shows a characteristic Chinese rig; yet in the image, there are clearly drawn quarter rudders, identical to those of ships from Southeast Asia or perhaps the Indian Ocean. A bowsprit seems to be drawn, and this is also unlikely to be attributable to the Chinese at the time. Another example is an image of the Japanese ship captured offshore in the Philippines in 1600 by the Dutch (Nagazumi 2001: 58). Despite having been labeled as a Japanese ship, the depicted ship has quarter rudders and square batten sails made of bamboo. These features are not characteristic of Japanese ships.

Such misperceptions by European recorders is probably due partly to the vagueness arising from the multiple nationalities present in a trade by then employing merchant ships and seamen from different countries. It might also have resulted from the artist's preconceptions about similarities between Southeast Asian and Chinese ships. Alternatively, perhaps such a misperception can be explained by the fact that to some degree the ship depicted shows hybrid features, integrating hull components from different regions.

There is clear evidence that Europeans recognized the existence of hybrid ships in Southeast Asia: such ships came to be widely known by the Spanish term *mestizo*, indicative of a hybrid. The fact that the Europeans described ships as mestizo is indirectly evidenced in the historical Japanese text titled *Zoho Kai Tsushoko* (Studies on the intercourse and trade with the Chinese and foreigners, 增補華夷通商考), written in 1708 by Joken Nishikawa (Ishii 1995b: 72–76). According to this text, the mestizo ships are explained as follows:

> Rigs and sails of the two masts of the ship are the same as those of a Chinese ship, and the depth of the hull as well. Its rudder is fitted with iron pintles that slip into iron gudgeons, and the stern is like those of ships from Fujian or Tanzhou. Structurally, these ships are also the same ships that used to sail from Nagasaki to Tenjiku, known as a vessel of mestizo construction. The large type of this ship could carry cargos of 2,000,000 *kin* [likely over a few hundred tons, but the conversion is uncertain], the middle-sized ships could carry 1,560,000 kin, and 1,230,000 kin could be loaded into the smaller ones. The ship has a bowsprit where there is a sprit sail, likewise observed on many Chinese ships voyaging overseas, also having top sails made of cotton, used for the sprit sail too [as distinct from the other sails, made of bamboo]. The length of the ship ranges from 15–16 to 20 ken [28–30 to 37 meters], as in ships from Siam [Thailand]. The upper part of the ship is painted with red terracotta color or simply painted with oil, while the lower part under the waterline is painted with substances made from lime and oil, which makes it appear to be painted white. (*Zoho Kai Tsushoko* [author's translation])

Apart from establishing that foreigners who visited Southeast Asian waters were cognizant of the hybridization, modern scholars have addressed the existence of hybrid ships in their own way. Needham (1971: 433–39) introduced an idea of affinities and hybrids of ship designs and construction, and he discussed an obsolete diffusion theory that locates diffusion in the spread of innovations from one focal point to peripheral zones simply by observing the similarities of features existing between two distant regions. This idea, developed by cultural anthropologists, was still dominant at the time of Needham's research. On the basis of such a view, early interaction between shipbuilders of Chinese ships and indigenous ships in Southeast Asia, yielding similar structures and construction methods, was proposed (Needham 1971: 438–39). According to Needham (1971: 438–39), the hybrid ship exhibits the Indo-China style; its development has been explained by the fact that a keel was perhaps the first feature to be adopted, and then the bulkheads were abandoned for frames. The *twaqo,* or *twakow,* used until recently in the waters of Singapore, serves as an example of such boats, showing similarities to Chinese ships from Fujian Province (Waters 1946). When looking at ethnological boat records, Needham noted similarities not only in the hull but also in the rigs and sails. Singapore's ships adopting rigs from distinctive Chinese batten sails has been highlighted as a possible case of hybridization. Yet Needham did not clearly explain innovative processes in rigging. He briefly proposed an adaptation theory mentioned by Jean Poujade: "Hulls tend to change more easily than rigs . . . [because] the shipwrights and the sail-maker are quite different people. Hulls . . . are sensitive to commercial contact; rigs, more closely associated with the tradition and life of the sea-faring folk, change only under political dominance" (Poujade, quoted in Needham 1971: 439). Furthermore, Needham indicated that Southeast Asia has been a focal place for hybridization for a long time, quoting Poujade's study regarding a ship inscribed on the wall relief of the temple of Bayon, built in the late twelfth or early thirteenth century. Poujade concluded that most of the particular features of the ship depicted in the relief show more similarities with Southeast Asian ships, but the hull is a hybrid, and the bulkheads coexist with a keel and a genuine stem and stern post (Needham 1971: 439).

As already mentioned, the idea of hybridization, specifically highlighting not rigs and ship configurations but integration of hull components, was put forward by Pierre-Yves Manguin and has become an established theory in interpreting shipbuilding traditions in Southeast Asia and East Asia (Manguin 1984, 1993). Following the identification of some characteristics in the hull of the early to mid-fifteenth-century Bukit Jakas site in Indonesia, Manguin (1993: 271) pointed out that several ship remains dating after the thirteenth century and archaeologically excavated in the

South China Sea had been constructed according to principles from two distinctive shipbuilding traditions.

An underwater archaeological excavation project by French scholars at a wreck site in Brunei waters (known as the Brunei shipwreck) has brought forward the idea of a hybrid ship to assess the details of the wrecked ship (L'Hour 2003, 2006). While the hull has not been identified, the ship carried a large cargo of ceramics similar to ceramic assemblages discovered with other excavated hybrid ships. The Lena Shoal shipwreck of the Ming dynasty (AD 1368–1644), found off the island of Busuanga in the Philippines, may be an interesting example: large numbers of Chinese ceramics were found at the site, which indicates the ship's involvement in maritime trade between China and Southeast Asia in the 1500s. The hull planking consists of double-layered planks that are edge-joined with wooden dowels (Goddio et al. 2002: 22–25). The hull is said to have been constructed based on bulkheads instead of frames (Goddio et al. 2002: 26). Although neither actual bulkheads nor frames have been identified, the use of bulkheads is said to be evidenced by the distribution of cargo, showing a rectangular-margined space (Goddio et al. 2002: 26). Numerous shipwrecks salvaged in Southeast Asian waters and dating from the sixteenth century onward have hull configurations similar to those of these two shipwrecks (Flecker 2007).

A hybrid ship in Southeast Asia is recognized as an integration of portions of the ship and hull components among people from outside the region, likely from southern China (Manguin 2010). The idea of hybridization of shipbuilding technologies was enunciated by modern scholars initially on the basis of ethnological boat studies and was later developed using historical and archaeological resources. In his early study of the hybrid ship, Manguin (1984) suggests that such hybridization could have occurred around the tenth century in the South China Sea. Further study is needed to elucidate the emergence and transitional processes of hybridization, especially through understanding of the historical dynamics of shipbuilding in the East Asian realm. Shipbuilding along the central and southern Chinese coasts has been outside the purview of mainstream traditional Chinese shipbuilding for a long time, being affected early by people from Southeast Asia and eventually by Persian and Arabic merchants as they also entered the maritime trade. A genuinely seaworthy ship may have originated in that area around the twelfth century, following the encounters of previous periods. The fact that the Europeans colonized the region and mentioned ships of mestizo construction indicates that technological integration had been substantially developed by the fifteenth century.

The advantages of integration were conceived outside the South China Sea as well. In the first half of the seventeenth century, a ship known as the red seal vessel (*shuinsen*) was used for long-distance maritime trade between Japan and Southeast Asia (Iwao 1958; Nagazumi 2001). The ship was defined as a historical Japanese merchant ship authorized by the Tokugawa Shogunate to conduct foreign trade, with a license having a red seal. Ships used in this trade were initially large vessels built in China; later, ships showing hybrid features started to be used, although where those ships were built is disputable—possibly in Japan but also possibly in Southeast Asia, perhaps Thailand. Representative configurations of red seal vessels are to be found in many Japanese iconographic resources depicting this type of ship (figure 1.1).

Hybridization is an important aspect of the formation of shipbuilding traditions. Through hybridization we can gain a better understanding of distinctive hull configurations, as evidenced in historical, iconographic, and archaeological resources. It is also important in gaining an appreciation of the dynamism of the regional shipbuilding tradition, the chronological milestones of which can be located by identifying structural hybridity with associated construction methods in excavated ships.

Regional Perspective on East Asian Shipbuilding

The existence of distinctive shipbuilding traditions has been noted by local researchers in each East Asian nation (Ishii 1957; Kim 1994; Xi et al. 2004; Yang 2009). These studies, however, have placed the historical transitions and epochs of ship development only within each nation's own chronology, in a way that explains the growth of endemic features in shipbuilding tradition as a linear evolutionary development. My emphasis is on the importance of a regional perspective, because hybridization in hull structure and construction methods hypothetically occurred at the regional level, where innovations arose not only from endogenous growth but also through exogenous factors. In this perspective, hybridization is attributed to technological innovations considered to have been factors

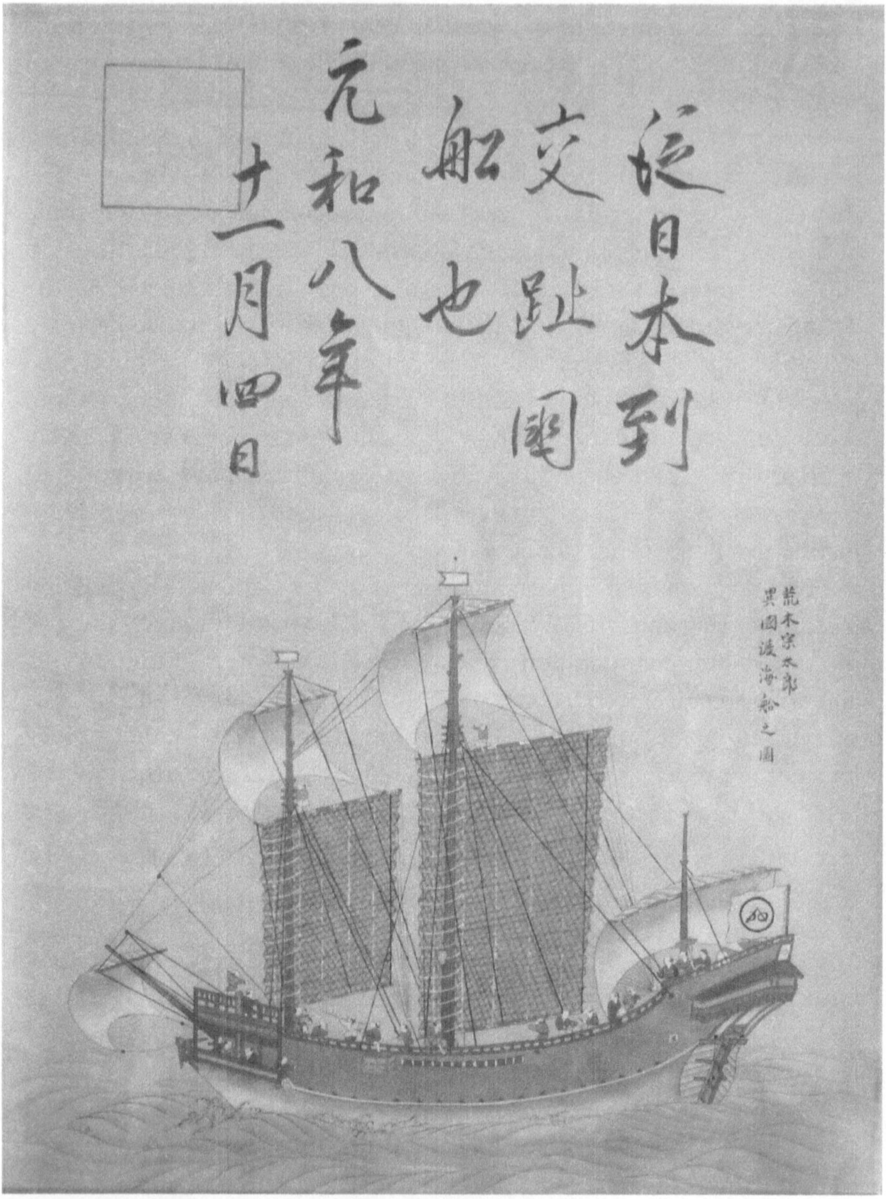

Figure 1.1. Historical scroll painting depicting a red seal ship known as an *araki bune*, sailing to Vietnam. (Courtesy of Museum of Maritime Science)

forming the regional shipbuilding traditions. Three segments of regional traditions are contrasted in the rest of this chapter: those for the Yellow Sea, the East China Sea, and the South China Sea.

Yellow Sea Area

The Yellow Sea shipbuilding tradition is primarily identified in ships having flat bottoms and operating in the northern waters of East Asia. They show two types of transverse components: beams and bulkheads. An example of the first type of early East Asian shipbuilding technology can be seen in the Korean coastal traders of the Goryeo dynasty, which have flat bottoms (see chapter 2). The hulls of nine Goryeo traders were archaeologically investigated offshore along the west coast of the southern peninsula, facing the Yellow Sea. The dates for the ships that exhibit this characteristic hull structure range from the eleventh to the fourteenth century, and this is perceived to be a long-term technology that consistently used beams, not bulkheads. The technology developed in the Korean Peninsula, represented by baulks as longitudinal structures and athwart beams as transverse structures, resulted in the construction of ships with sufficient seaworthiness to sail in East Asian waters, especially under the environmental conditions of the Yellow Sea. This could have facilitated the dominant role of merchants who originated in the Korean Peninsula in maritime trade between that peninsula and Japan before the tenth century. In this context, Japan's shipbuilding tradition of constructing wooden hulls with a flat bottom and transverse beams is addressed as a possible early interaction with shipbuilding in Korea (see chapter 2).

Early Chinese ships, in contrast, used bulkheads, as is evident in excavated ships dating to the tenth century, during the Tang dynasty (Xi et al. 2004; Kimura 2010b). Two Tang dynasty ships, probably used for inland water transportation, were discovered near rivers in Jiangsu Province (see chapter 3). The province is located at the mouth of the Chan (Yangtze) River, and another large river, the Huai River, flows through it. Historically the area was part of the construction of the Grand Canal. This setting promoted utilization of inland water transportation, for which watercraft having a flat bottom were preferable. The region faces the Yellow Sea, a shallow sea where people could still use the flat-bottomed ships originally designed as riverine ships. Jiangsu Province and the adjacent Zhejiang Province to the south had ports for ships conducting distant voyages before the tenth century. Some envoy ships, for example, regularly sailed between the Chinese continent and Japan during the Tang dynasty (Reischauer 1940). To date, however, no ships have been identified that are known to have been specifically designed for seafaring and date to that period. Therefore, we can only deduce that the structure of those ships that people used to sail into the open waters could be identical to that of riverine ships. For example, in much later periods the *sha-chuan* (sand ships), as represented by the Hangzhou Bay trader or the Nanjing ship, were used for both riverine and oceangoing purposes in the Yellow Sea and the northern East China Sea (see chapter 3). Hence, the technology of the early stage of building oceangoing vessels along the northern coast of China can be regarded as similar to technology in use for the construction of riverine boats at the same time.

The flat-bottomed ships, as represented by the Goryeo dynasty coastal traders and the early northern Chinese ships for riverine transportation, constitute an ancestral form of seagoing ships. The definition of the Yellow Sea shipbuilding tradition thus encompasses two distinctive shipbuilding technologies from the Korean Peninsula and northern China. Technological interactions and mutual influences that might have occurred between these two regions are still speculative. Nevertheless, by looking at archaeological evidence, one can argue for the high possibility of mutual influences among different shipbuilding traditions in the Yellow Sea area. The discovery at Penglai of later ships of the Ming dynasty (AD 1368–1644 [see map 1.1]), however, is indicative of technological hybridization between shipbuilding in the East China Sea and in the Korean Peninsula (see chapter 7).

East China Sea Area

Two detailed case studies of archaeologically excavated ships originally from the coasts of the East China Sea reveal the existence of a distinctive shipbuilding tradition (see chapters 4 and 5). The technological innovations identified on these excavated ships suggest an East China Sea shipbuilding tradition that generally shows either V-shaped or U-shaped bottoms, a keel and bulkheads, and multiple-layered hull planking. Ships built according to this tradition came to be used in seaborne activities within and beyond East Asia and Southeast Asia.

In middle and southern China, early shipbuilding traditions before the tenth century can be only vaguely presumed. Contemporary historical records of Chinese Buddhist pilgrims who departed from the southern Chinese coast indicate the use of Southeast Asian indigenous ships to cross the South China Sea (Wang 1958; Hall 1985). Before the tenth century, some cities from south of Hangzhou Bay (around Ningbo) to southern China (Guangzhou), facing the East China Sea and South China Sea, were the focal area of trade by foreign merchants from Persia and Arabia (Kuwabara 1989). The influence of shipbuilding traditions from outside East Asia on the existing shipbuilding industries along the central and southern Chinese coasts is, however, unclear.

Technological development of the East China Sea shipbuilding tradition clearly occurred during the Song dynasty (AD 960–1279) and the maturity of shipping business during the period, continuing in the Yuan dynasty (AD 1271–1368). Political entities relying on naval power were formed in the late Song and early Yuan eras (Lo 1955). Shipyards serving long-distance traders emerged around the areas southward from Hangzhou Bay and coasts around the Taiwan Straits. Two archaeologically excavated ships, the late thirteenth-century Quanzhou ship and the early fourteenth-century Shinan shipwreck, originated in these regions. They are the most pertinent archaeological resources for addressing in detail the structural characteristics of the East China Sea shipbuilding tradition. Their hulls represent a standardization of the structure for oceangoing ships, which is interpreted as one stage of the completion of technological integration or clustering of hull components in East Asian shipbuilding traditions.

The keel and hull planking contribute synergistically to longitudinal strength. In general the keel is composed of three members, and thick garboards are fixed. The hull planking that extends from the garboards forms a steep deadrise angle. Hull planking is edge-fastened by skew nailing with rabbeted seams. The system of multiple-layered hull planking is standardized to protect the inner planks and in some cases to enhance the longitudinal strength of the plank shell. The bulkheads provide transverse strength. Large half frames are attached to secure the bulkhead planks, fixing them into the hull planking with brackets.

Ship construction materials including ship nails and timbers are highlighted later in the volume through analysis of samples from the Quanzhou ship and the Shinan shipwreck and with reference to data from the Takashima underwater site in Japan (see chapter 6). That site gives an absolute date related to the invasion of Japan by Kublai Khan of the Mongol Empire (during the Yuan dynasty) in 1281 and is an important complementary resource to the two ship remains mentioned above. The remains of an iron nail collected from a hull plank of the Shinan shipwreck are described in detail to address manufacturing procedures, which are similar to the processes used for some ships of Kublai Khan's fleet. A timber sample recently collected from the Quanzhou ship was used for wood species identification. Results are presented for comparison with the assemblages of wood species of the other excavated ships in East Asia (see chapter 6).

A few other ships more recently excavated in Chinese waters and similar to oceangoing traders of the East China Sea are reviewed in chapter 7, which also provides a perspective on the existence of another shipbuilding tradition in the East China Sea area. While the assessment of ships built in this tradition is not as comprehensive as for the first type of East China Sea ships, it is recognized as distinctive in that the hull has a bulkhead structure yet shows a U shape in cross section and has single-layered hull planking. This second type of East China Sea ship appears to reflect interaction with the shipbuilding technologies developed along the northern coast of China, including the Yellow Sea region.

South China Sea Area

The concept of the South China Sea shipbuilding tradition was built on observation of ship remains with distinctive features, showing that the hull configuration is identical to that of the Chinese shipbuilding tradition but that some construction methods (for example, the use of edge-joined planking with wooden dowels) more likely derive from Southeast Asian traditions. The configurations of early Southeast Asian seagoing ships are illustrated in bas-reliefs of monuments and in the pilgrimage records of Chinese monks, but these sources lack detailed information about physical structure and construction methods. My review notes an increase in the discovery of ship remains showing features consistent with Southeast Asian shipbuilding traditions spread over the nations of both mainland and insular Southeast Asia (see chapter 8). The fact that the key boatbuilding innovations were disseminated widely and were dominant along the coasts of Southeast Asia supports the hypothesis of transitional processes of hybridization occurring in the South China Sea area, whereby the shipwrights employing the dominant technology endemic to the region could have been receptive to an exogenous shipbuilding tradition. The South China Sea shipbuilding tradition is exemplified by a type of hybrid ship integrating East China Sea shipbuilding traditions into Southeast Asian (Austronesian) traditions. Recent discoveries of hybridized hulls in the Philippines and Thailand buttress this hypothetical idea (see chapter 8). Thus, while the presence of hybrid ships has been acknowledged in previous research, the present study traces the formation of such ships chronologically as they evolved from and related to adjacent shipbuilding traditions, including those of the Yellow Sea and East China Sea areas.

2

Shipbuilding in Early East Asia

The strait that lies between the southern Korean Peninsula and Japan historically served as the most significant access route between the two regions. Early Japanese envoys perceived passing through this strait to the Korean Peninsula and the Yellow Sea to be safe, and during the ninth century this route was used by other East Asian merchants to reach Japan, in particular (Yamauchi 2008). Seafarers along the coast of the Korean Peninsula had a high profile in East Asian waters ahead of the rise of Chinese merchants. Even before the early ninth century, merchants from Silla (57 BC–AD 935), one of the Three Kingdoms of ancient Korea, initially played a significant role in trade with Japan and in ship transportation for the Japanese to reach China. The activities of Silla merchants appear in the historical record (Matsunami 2006; Enomoto 2008) of early international maritime trade in East Asia. The detailed configuration and structure of Silla ships are unknown, given the lack of concrete evidence. However, by the tenth century the traders had probably developed flat-bottomed ships that had sturdy transverse beams in their interior structure. These features can be deduced from the features of archaeologically investigated coastal traders dating to the Goryeo dynasty (AD 918–1392).

While a broader range of mutual cultural influence between Korean and Japan occurred early, there has been almost no discussion of interaction between the two regions in shipbuilding. Japanese researchers have made inroads through study of the evolutionary processes of traditional boats in Japan and have offered an explanation of chronological development of the structure of indigenous ships, presented as a linear development theory to elucidate how traditional types of watercraft descended from dugout canoes. The growing number of excavated ships of the Goryeo dynasty in Korea calls such a linear development theory into question. The hulls of the Goryeo traders show consistency in construction principles and represent one of the early shipbuilding traditions for coastal traders in northern East Asia. The formation of the long-term tradition could be idiosyncratic to the Korean Peninsula, but its early impact on the Japanese shipbuilding tradition can be presumed.

Reorienting the Existing Hull Evolution Theory

In Japan, the remains of dugout canoes are what have predominantly been found, leading to the above-mentioned discussion of a linear process from dugout canoes to indigenous coastal traders as a focal theory of the structural development of ships (Ishii 1957, 1995a). Kenji Ishii (1957: 6) pointed out that more than 70 dugout canoes had been discovered in Japan by the 1950s. According to a recent report, the number of dugout remains discovered that date to the prehistoric period is 117 from all over Japan (Shiga Prefectural Association for Cultural Heritage 2006), and this number increases to more than 200 when dugout remains dating to later periods are included.

Most Japanese scholars in nautical history support the idea of chronological development of traditional watercraft in three consecutive stages: from dugout canoes to planked-up dugouts and then to ships having a more seaworthy structure (Ishii 1957; Shimizu 1975; Matsuki 1988; Adachi 1998; Okita 2000a). The planked-up dugout is generally considered to have consisted of a dugout base as the bottom structure with strakes added on its sides. Actual archaeological remains of planked-up dugouts are few, aside from the notable discovery of one at the Kyuhoji site in Osaka. The configuration of the planked-up dugout could have been as represented in small earthenware ship figurines, and one such archaeologically discovered figurine was used to reconstruct a planked-up dugout (figure 2.1). Any type of boat transitional between the dugout canoe and seaworthy ships is a candidate for interpretation as a planked-up dugout, and this has become a significant part of the evolutionary theory of traditional boat structure. Detailed development of planked-up dugouts, however, has not been insightfully argued beyond the structural configuration. Apart from an extensive study of dugout canoes by Akiko Deguchi (1995), we lack a detailed assessment of how the transition in construction methods developed alongside the structural development: important aspects not fully addressed include the sequence of construction, enhancement of longitudinal and transverse strength, and associated fastening methods. Deguchi's (1995: 52–53) study, which is based on ethnological

Figure 2.1. Reconstructed planked-up dugout canoe at the Osaka Maritime Museum; the lower part consists of a hollowed-out log. (Photo by author)

and archaeological data, provides insights into the construction methods of Japan's dugout canoes, compared to some ancient dugout canoes discovered in China (Dai 1985; Xu 1985; Deguchi 1995; Xi 2000; Xi et al. 2004).

Deguchi's research offers a regional perspective on the use of primitive watercraft in East Asia, including not only dugout canoes but also rafts. Rafts are regarded as another important primitive watercraft relevant to early shipbuilding development in East Asia (Hornell 1934; Donnelly 1936; Needham 1971; Worcester 1971; Greenhill 1976). Seán McGrail (2001) has mentioned that construction materials of East Asian rafts changed at around 33 degrees north latitude: farther north, rafts tended to be constructed from logs, such as cedar, pine, and *Paulownia* sp., whereas farther south, bamboo was a common raft construction material. Deguchi (1995: 54–55) has focused on unique fastening methods of the log rafts from Japan and Korea, in which two holes transversely penetrate the timbers of the rafts forward and aft, and through these holes pass wooden bars that are fixed by wooden nails perpendicularly driven from the top surface of the timbers. According to Deguchi, the fastening methods observed on the modern log rafts from Japan and Korea are similar to those of the Goryeo dynasty coastal traders.

Whether a consecutive progression occurred from techniques for the construction of rafts to technologies to build the Goryeo traders has not been verified. The prominence of raft construction techniques in East Asia must be carefully argued, because of the paucity of archaeological evidence, especially in Japan. The paucity of data has meant that descendent technologies can be regarded only as locally developed techniques. Planked-up dugouts could have been a primary part of a linear development from primitive watercraft to seagoing ships. The endogenous growth of technology in Japan is considerable in shipbuilding. But exogenous factors deriving from interaction with other regions have largely been overlooked in the current development theory.

Flat-Bottomed Ship and Beams

As already mentioned, Silla merchants based on the Korean Peninsula were the most active seafarers in early East Asian maritime history. During the late eighth century, merchants from Silla voyaged to visit Dazaifu, an administrative center in southern Japan responsible for foreign affairs. Jang Bogo, a prominent maritime figure from Silla, visited Dazaifu in 824 (Matsunami 2006). Details of the Silla ships used by the traders have not been depicted in historical documents. The ships' seaworthiness, however, is partly illustrated by the fact that in 837 the central government of the time in Japan gave Dazaifu orders to construct Silla-type ships (Matsunami 2006). In 840 a local bureaucrat of Tsushima Island, located between southern Japan and the southern end of the Korean Peninsula, requested that Dazaifu provide one of the better-constructed Silla ships because of the high frequency of wrecking of the native Japanese ships (Matsunami 2006). Some envoys from Japan to China appear to have preferred to use Silla ships as well.

The following section describes the structural details of the Goryeo coastal traders. The long-term continuity of structural design during the Goryeo dynasty and much later periods suggests that the same construction methods might have been used for earlier Korean ships. The absence of bulkheads and the use of fastening methods relying on wooden fastenings could be representative features of Korean shipbuilding.

In Japan, compared to Korea, the archaeological discovery of watercraft has been limited. Iconographies, ethnographic study of boats, and historical texts have been recognized as the best clues available for the study of Japanese seagoing ships. Although using written and drawn information makes detailed examination of hull structure and construction techniques impossible, Ishii (1957: 16–17) indicates that the development of flat-bottomed inland watercraft occurred by the ninth century. Furthermore, he mentions the use of flat-bottomed ships to conduct seagoing voyages to China by the fourteenth century, citing the sixteenth-century Chinese text *Chou Hai Tu Bian* (Illustrated seaboard strategy, 籌海図編). The final part of this section presents an archaeological assessment of Japanese coastal traders, based on the limited available data, to address their featuring of a flat bottom and transverse beams identical to those in the Korean shipbuilding tradition.

Hulls Excavated in Korea

Korean shipbuilding shows a long-term tradition, generated from technologies to construct ships used for sailing along the coast of the Korean Peninsula and for seagoing in the Yellow Sea. This specialization formed a suite of idiosyncratic features in the hulls of those ships, with many similarities between the late eleventh-century coastal trader and the early twentieth-century coastal fishing ship (figure 2.2). In general, the traditional Korean ships have a flat bottom with the transom bow and stern forming a box-shaped hull. The planks are edge-fastened with rabbeted seams, making the hull planking appear to be clinker-built. The use of wooden fastenings is a common feature. Beams are used as transverse components. An archaeological discovery of a Korean coastal trader

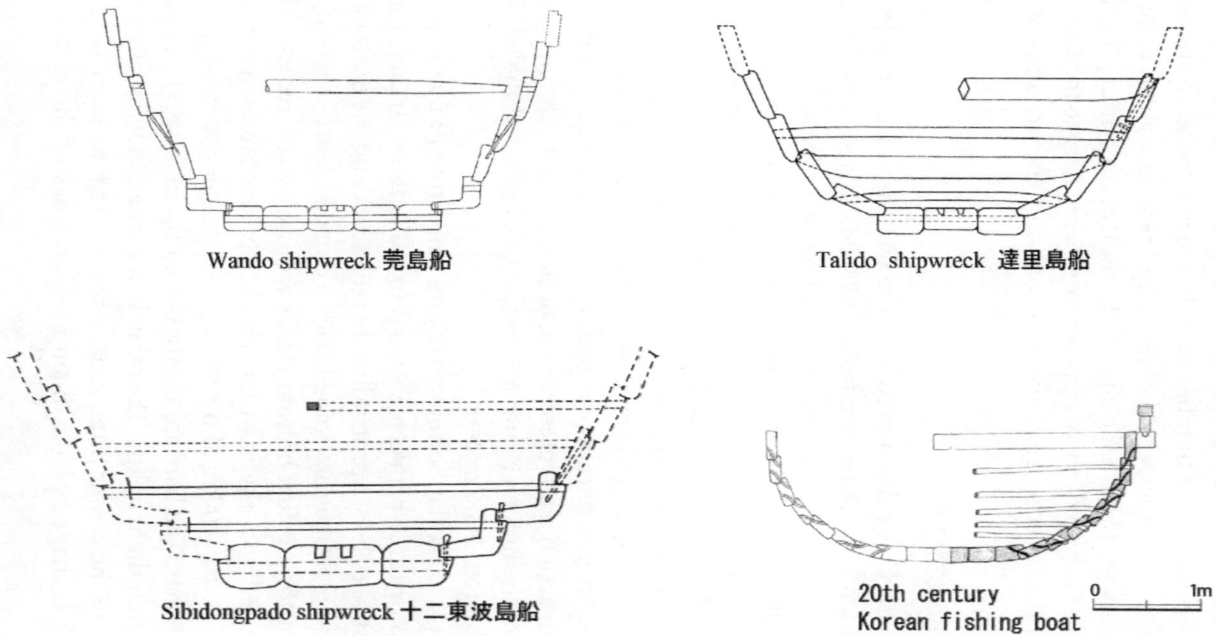

Figure 2.2. Cross sections of the Wando and Sibidongpado ships (*left*, late eleventh to early twelfth century) and the Talido ship (*top right*, thirteenth to fourteenth century), compared with a modern Korean fishing boat. (After National Maritime Museum of Korea 2008b)

dating back to the early Goryeo dynasty was reported early by Jeremy Green (1983b) and has since been reviewed again with the addition of recent discoveries and conservation work (Sasaki and Lee 2010; Moon 2011; Yang 2011). Underwater archaeology in Korea has been growing rapidly, and the National Research Institute of Maritime Cultural Heritage has identified more underwater sites on the west coast of the central Korean Peninsula; the number of hull remains found from the 1980s to 2011 is nine. The time period of these hull remains ranges from the eleventh century for the Wando ship to the fourteenth century for the Anjwa ship. These Goryeo dynasty ships are compared here to facilitate a critical review of the chronological evolution of hull structure and construction methods.

Wando Ship (莞島船)

The shipwreck was discovered offshore of Wando Island in Jeolla Province, South Korea, and excavated in 1983–84. This was the first shipwreck of Korean origin to be discovered, and more important, it is among the earliest shipwrecks recovered from East Asian waters. A few brief reports are available in non-Korean-language resources (Green and Kim 1989; Kim 1991; McGrail 2001: 361–63). The presence of more than 30,000 pieces of Goryeo celadon suggests that the ship dates to the second half of the eleventh century (National Maritime Museum of Korea 2005a). The wood species identification suggests a Korean origin, and the structure and construction are representative of the Korean tradition. The remaining part of the hull measures 6.5 meters in length, and the widest part of the hull bottom is 1.6 meters.

The hull has a flat bottom structure consisting of five strakes (figure 2.3). Two or three planks are joined to make these strakes. Butts of the median strake show a protruding-tongue scarf, and the butts of the other four bottom strakes are joined with a stepped scarf. Around the midsection of the median bottom strake are two square recesses that could be associated with the mast tabernacle. Six long wooden bars penetrate the three center strakes of the five bottom strakes athwartships, fastening them together. The bars are neatly slotted through apertures cut in each strake to join them tightly. The other two strakes, individually joined onto each side of the center three strakes, are fixed by six mortise and tenons. There are L-shaped planks fitted on both sides of the bottom, defined

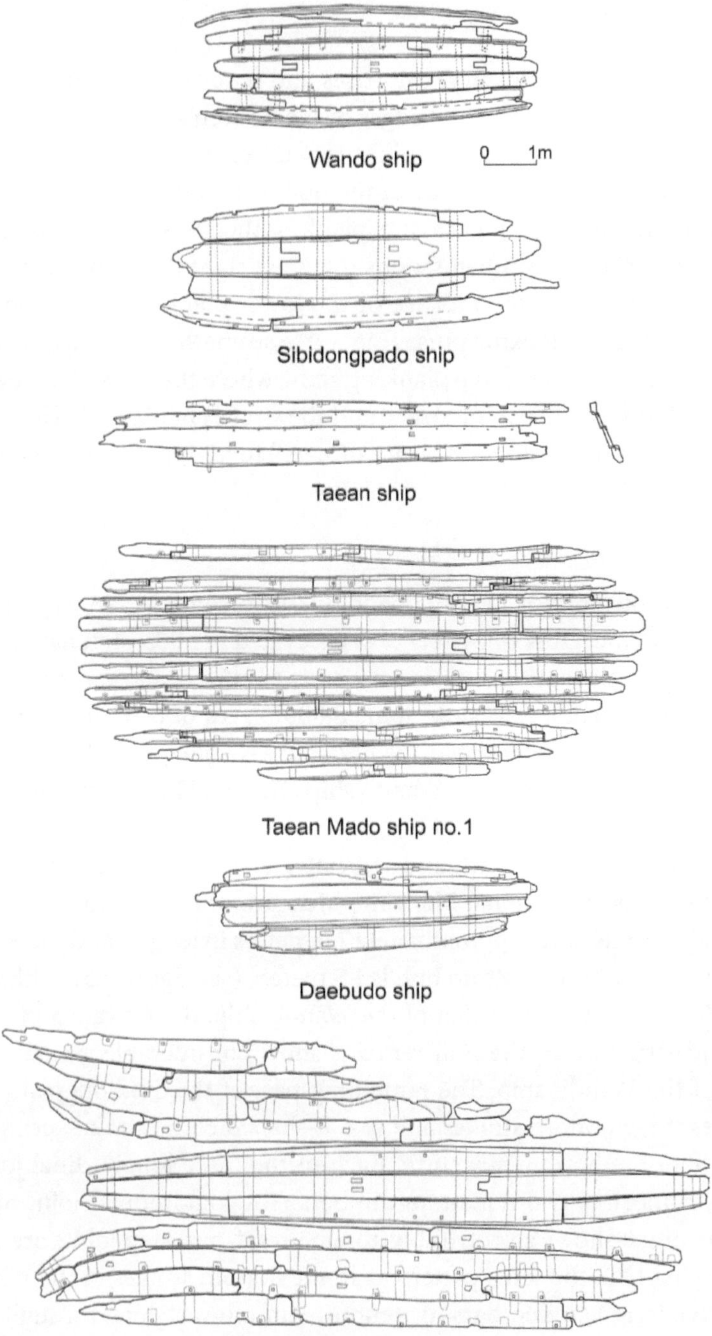

Figure 2.3. Plans of the hull remains of Goryeo traders. (After, from top: Green 1989; National Maritime Museum of Korea 2005b; National Maritime Museum of Korea 2009; National Research Institute of Maritime Cultural Heritage 2010; National Maritime Museum of Korea 2008a; and courtesy of the National Maritime Museum of Korea)

as chine strakes (see glossary). Mortises and tenons fasten the L-shaped chine planks onto the rabbeted edges of the bottom strakes.

Five side strakes above the chines remain on one side and four side strakes remain on the other side of the hull. A rabbet is cut on the outer part of the upper edge of each side plank of the strakes. The planks are fastened together by wooden tenons that pass right through the mortise plank and down into the mortise of the lower plank. These tenons are secured by wooden locking pins. Transverse strength is provided by multiple beams. Apertures in the planking show where the ends of the beams protruded through the chine strake and the strakes of the hull's sides. The original length of the ship has been estimated to be approximately nine to ten meters.

Sibidongpado Ship (十二東波島船)

Discovered offshore of Sibidongpado, Okdo-myeon, in Gunsan city in Jeonbuk Province, this shipwreck was excavated in 2003–4. It dated from the end of the eleventh century to the early twelfth century, according to the analysis of artifacts, mainly some 8,000 pieces of ceramics. A report published more recently dated the ceramic assemblage of this ship to a period earlier than that of the Wando ship (National Research Institute of Maritime Cultural Heritage 2009).

The details of hull remains and associated artifacts are presented in the excavation report (National Maritime Museum of Korea 2005b). The remaining hull measures approximately 7.0 meters in length, and the widest remaining part of the bottom hull is 1.8 meters (see figure 2.3). Although the hull is less intact than that of the Wando ship, the overall configuration and structure of the hull remains show considerable similarity to those of the Wando ship. The remaining part of the Sibidongpado ship includes three bottom strakes and two L-shaped chine strakes sitting on one side of the bottom structure. Each of the three longitudinal joined strakes comprises two remaining timbers. The longitudinal joint of the median strake shows a protruding-tongue scarf, and step joints are used in the other two, the same pattern as in the bottom strakes of the Wando ship. Five long wooden bars, or tenons, run athwartships through rectangular mortises cut through the three bottom strakes. The upper edges of the planks of the outer strakes are rabbeted to receive L-shaped chine planks. The most distinctive feature of the Sibidongpado ship compared to the Wando ship is the double L-shaped chine strakes: the lower chine strake is placed on the bottom strakes, and the upper chine strake sits on the rabbeted edge of the lower one. The presence of two beams is inferred from the square mortises in the lower chine strake into which they were fitted.

Several timbers detached from the ship, including part of the square bow transom, were scattered around the remaining part of the hull. Although the bow remains are substantially degraded, three planks show evidence of wooden tenons used to edge-join them. Of the three planks, the best preserved measures 1.6 meters in length. The edges of the bow transom are recessed and form steps where the clinker-arranged hull planking could meet. The other detached timbers include a piece of a hull plank and timbers possibly associated with an anchor windlass. The original length of the ship has been estimated as approximately 14.0 meters.

Taean Ship (泰安船)

The shipwreck was identified in 2007 offshore of Daeseom Island in Taean through the discovery of a celadon bowl by a local fisherman. During the excavation in 2007–8 by the National Maritime Museum, more than 23,000 artifacts were recovered, including six ship timbers, two large stone anchor stocks (or anchor weights), pieces of a windlass, wooden strips, metal and earthen wares, stone bricks, bones, and 20,000 celadon pieces and products. The significance of this shipwreck has been recognized because of the quantity and quality of celadon products, some probably intended for the royal court; the shipwreck is known as the Goryeo Celadon Treasure Ship. Details of the recovered artifacts and hull are well documented (National Research Institute of Maritime Cultural Heritage 2009). The ship is presumed to have sunk during a voyage from Gangjin, at the tip of Korean Peninsula, to Kaesong (Gaeseong), the ancient capital of the Goryeo dynasty, in the twelfth century.

All of the six identified ship timbers are hull planks, constituting parts of four side strakes measuring 8.5 meters long and 1.5 meters wide (see figure 2.3). The method of plank fastening in the Taean ship differs from that of the preceding two ships. The wooden fastenings directly secure the seams of the planks; they are driven diagonally from the lower part of the upper plank to the upper part of the lower plank, like the nailing of Chinese ships. Locking pins appear to have secured these wooden fastenings.

The fastening method is still used in conjunction with rabbeted seams forming clinker steps. The lower and upper strakes, identified from the pattern of the fastening and clinker joint, are thicker than the other two (see figure 2.3). The remains of these thicker strakes, or wales, consist of two timbers joined with a stepped scarf. The two stepped scarfs are vertically aligned in the presumed configuration of the planking. All these side strakes have square holes probably associated with the use of beams.

Two large stones (115 kilograms and 70.5 kilograms in weight) that were part of anchors or used as anchor weights have been identified. Grooves or recesses on their sides are believed to have been for the anchor ropes. The estimated length of the original hull is more than twenty meters.

Taean Mado Ships (泰安馬道船) *Numbers 1–3*

The waters offshore of the Taean Peninsula in Chungcheongnam Province were an important sea route during the Goryeo dynasty. Coastal traders sailing along the west coast of southern Korea between north and south transported goods to the Goryeo capital, Kaesong (Gaeseong), by passing through the Mado waters in Taean. The area, however, is notorious for wrecking because of changeable weather, complex currents, and the hazardous combination of muddy water and submerged rocks. Potential for the discovery of shipwrecks in the area had been presumed, but the identification of shipwreck sites did not occur until the seabed underwent changes related to the development of the surrounding harbor area. An underwater survey in 2008–9 identified sixty-eight stone anchors (including stone stocks) and more than 670 pieces of Korean and Chinese ceramics (Kim and Moon 2011; National Research Institute of Maritime Cultural Heritage 2011a). In addition, the three hull remains found during the survey were fully excavated. Archaeological reports of excavations of two of the hull remains had been produced by 2011 (National Research Institute of Maritime Cultural Heritage 2010, 2011b).

Taean Mado ship no. 1 was initially found in 2008 and was excavated during the following year. The main items among the excavated artifacts are approximately three hundred pieces of Goryeo celadon and a lot of coal and grain, the major commodities carried by the ship and indicated in associated tags and labels (Cha 2011). Korean scholars gave the wreck an absolute date: based on ink writing on the bamboo tags and labels, the wrecking probably occurred in 1208 on the way to Kaesong.

The hull is approximately 10.8 meters in length and 3.7 meters in width. Forty pieces of timbers related to the hull remains were taken from the waters and placed in a conservation laboratory of the National Research Institute in Taean. The hull has a wider flat bottom than that of the other early Goryeo traders, consisting of seven strakes (see figure 2.3). Four longitudinally joined timber remains compose the individual bottom strake. The bottom hull has a gentle curvature upward toward the bow and stern. Butts of the median strake show a protruding-tongue scarf, and the strakes adjoining the median strake are joined simply with a butt joint. These three center strakes are fixed together by ten long wooden bars penetrating them athwartships. Another two strakes are attached to each side by mortises and tenons, and the longitudinal butts are joined mostly with a stepped scarf. The fundamental bottom hull structure and the fastening methods of the timbers are almost identical to those of the Wando ship, but the chine timbers have a distinctive configuration. The Taean Mado ship does not use chine timbers having an L-shaped cross section. The chine consists of robust timbers chamfered roughly, but its interior has been carefully cut to have a flat surface, so that it shows as a semicircle in cross section. Three to four timbers remain in each chine. The upper edges of the chines are grooved to join upper planks, and the same applies to two strakes remaining in one side of the hull. Wooden nails are used as fastenings for the planking. Thinner planks could have been used in the upper strakes; they are missing. Notably, part of the bow transom remains, consisting of seven vertically joined timbers. The edges of the exterior sides of the timbers are chamfered to meet the clinker-arranged hull planking. The median strake has two square recesses (measuring 150 × 80 × 70 millimeters), probably related to a mast structure. The excavation reports indicate that living space for sailors could have been located in the middle part of the ship around the mast, based on the concentration there of the remnants of cooking facilities and daily utensils.

According to the excavation report (National Research Institute of Maritime Cultural Heritage 2011b), Taean Mado ship no. 2 was found in waters nine hundred meters east of Taean Mado ship no. 1. The full excavation, conducted in 2010, led to the discovery of well-preserved artifacts, consisting of more than four hundred objects, including Goryeo celadon from the Buan kiln and grain identified as the major cargo of the ship. The discovery of many bamboo tags and wooden slips associated with the cargo was significant, identifying the use of the celadon jars as containers for sesame oil and honey and dating the wrecking as the early thirteenth century.

The remaining part of the hull, comprising forty-one timbers, measures 12.6 meters in length and 4.4 meters in width. The fundamental hull structure is identical to that of Taean Mado ship no. 1; ship no. 2 has a flat bottom consisting of seven strakes and robust chine timbers. The number of remaining planks is greater than in ship no. 1: one strake on the port side and four strakes on the starboard side (these strakes and the chine strakes are all considered together as side planks in the report). Korean researchers have conducted three-dimensional modeling of each timber to integrate them into the entire hull. The overall hull shape shows the greatest breadth amidships well forward, with the bow transom bow at least 2 meters wide and the stern transom 1.5 meters wide. Excavation of Taean Mado ship no. 3 was completed in 2011, and structural features similar to those of ships nos. 1 and 2 have been confirmed.

Daebudo Ship (大阜島)

The ship remains were discovered in the intertidal zone of Daebu Island, Ansan City, in Gyeonggi Province, and were excavated in 2006. The results of the excavation are documented (National Maritime Museum of Korea 2008a). Ceramics discovered date the shipwreck from the late twelfth century to the early thirteenth century. Most of the hull is missing; only three bottom strakes and a chine strake remain. The remaining part of the hull measures 6.6 meters in length and 1.4 meters at the widest remaining part (see figure 2.3). The bottom of the hull was originally formed by five bottom strakes, but only three sets of bottom planks remain. Three long wooden bars penetrate the bottom strakes athwartships, and there are also several square mortises where shorter tenons or bars were fitted. The remains of the center bottom strake consist of only one timber with two recesses, possibly for a mast structure on the top, the same feature as in the Wando and Sibidongpado ships. Each of the other two bottom strakes consists of two longitudinally joined planks. The joint of the one next to the center bottom strake appears to be a butt joint, in contrast with the other bottom strake that holds the chine strake, which has a scarf joint. All these bottom strakes have a slight curvature upward toward the bow and stern, the same configuration as in the Wando and Sibidongpado ships. The chine strake consists of two planks. The planks of the chine strake were said to have been originally cut to an L shape in cross section, but degradation of the timber has removed the main part of the plank that would clarify the original configuration of the cross section (National Maritime Museum of Korea 2008a). The chine strake is joined to the bottom plank by mortises and tenons.

Talido (Dalido) Ship (達里島船)

The ship remains were found in 1995 in the intertidal zone on the shore of Talido (Dalido) Island in Mokpo, Jeonranam Province. The ship is dated to the thirteenth or fourteenth century based on radiocarbon analysis of its timber (BP 730±57). No artifacts are associated with the hull. The remaining part of the hull includes three bottom strakes and four side strakes on both sides. The remains measure approximately 10.5 meters in length and 2.7 meters at their widest. The median bottom strake consists of two planks joined by a tongue scarf. It has two square recesses, presumably associated with a mast structure. As the cross section shows, the median bottom strake is thinner than the other bottom strakes (see figure 2.2). The other bottom strakes, placed on either side of the median bottom strake, consist of three planks each, and they are lap-joined. These bottom strakes are slightly rockered and tapered toward the bow and stern, the same as in the other ships. The bottom strakes are joined by long wooden bars or tenons athwartships through mortises. Twelve of these mortises are observable. The absence of L-shaped chine planks can be confirmed, and the chine timbers are thick strakes. Three strakes above these chine strakes consist of three planks longitudinally joined by step joints, which are secured with tenons. For edge-fastening the hull planking, tenons are driven through rabbets cut in the upper edge of the upper plank to the lower plank, and pegs are used to lock the tenons. This fastening is used for the connection of the chine strakes and bottom planking as well.

Six beams have been identified, and these are distinctive in the way they are fixed into the hull planking: the thinner beams are slotted into square apertures cut in the hull planks, while the thicker beams are fixed into recesses in the inner surface of the hull planks. Four timbers of the stern transom remain. Although they are degraded, the third plank appears to have tenons on the side for the joint with the hull planking. The lowest side planks at one end of the hull appear to have recesses, which might have been cut for the transom stern (or bow).

Anjwa Ship (安佐船)

Discovered at Guemsan, Anjwa-myeon, in Shinan district in 2005, the ship remains were excavated the same year. The results of the excavation are available in an archaeological report (National Maritime Museum of Korea 2006c). Although the upper structure of the hull is missing, the remaining parts are well preserved, including three bottom strakes, two side strakes on the port side, seven starboard side strakes, part of the stern structure, a few beams, and a portion of what was possibly a sculling oar. A few celadon sherds were discovered beneath the hull. They provide an approximate date for the ship, which was dated to the late fourteenth century. The remaining part of the hull measures 14.5 meters long and 6.1 meters wide.

The median bottom strake consists of two large timbers joined by a tongued half-lap scarf, which differs from the joinery of the center plank in the other ships (figure 2.4). Fourteen mortises are cut through to receive long wooden bars to join all three bottom strakes. The bars appear to have been secured by locking pegs. The median bottom strake has two square recesses for a mast structure near the midpoint. The two other bottom strakes consist of two planks, joined with lap joints. All the bottom strakes have a slight rocker and are tapered toward the bow and stern. All the bottom strakes also have a recess near their end at the bow, forming a groove into which a transom bow was slotted. Planks of the first strakes building the sides up from the chines are thicker than the other hull planks and comprise four planks with step joints. The planks of this first side strake in the bow are distinctive in that they are triangular in cross section (figure 2.4). The aft planks of this strake have grooves or recesses to secure a plank of the transom stern. Four planks are used for each of the intact strakes, with step joints fastened by dowels (figure 2.4). The edge-fastening of the hull planking is by the same method as that of the Wando, Talido, and Sibidongpado ships (figure 2.4).

Five beams have been identified. A thicker beam was fixed athwartships into the second strake. It has two square recesses cut through its middle parts, which presumably allowed longitudinal timbers to run through it from bow to stern. Thin beams are subclassified into two types based on the different ways of fixing them into the hull planking. The end of the first type penetrates through the hull planks and is secured by wedges; the end of the second type is hooked on recesses cut into the inner surface

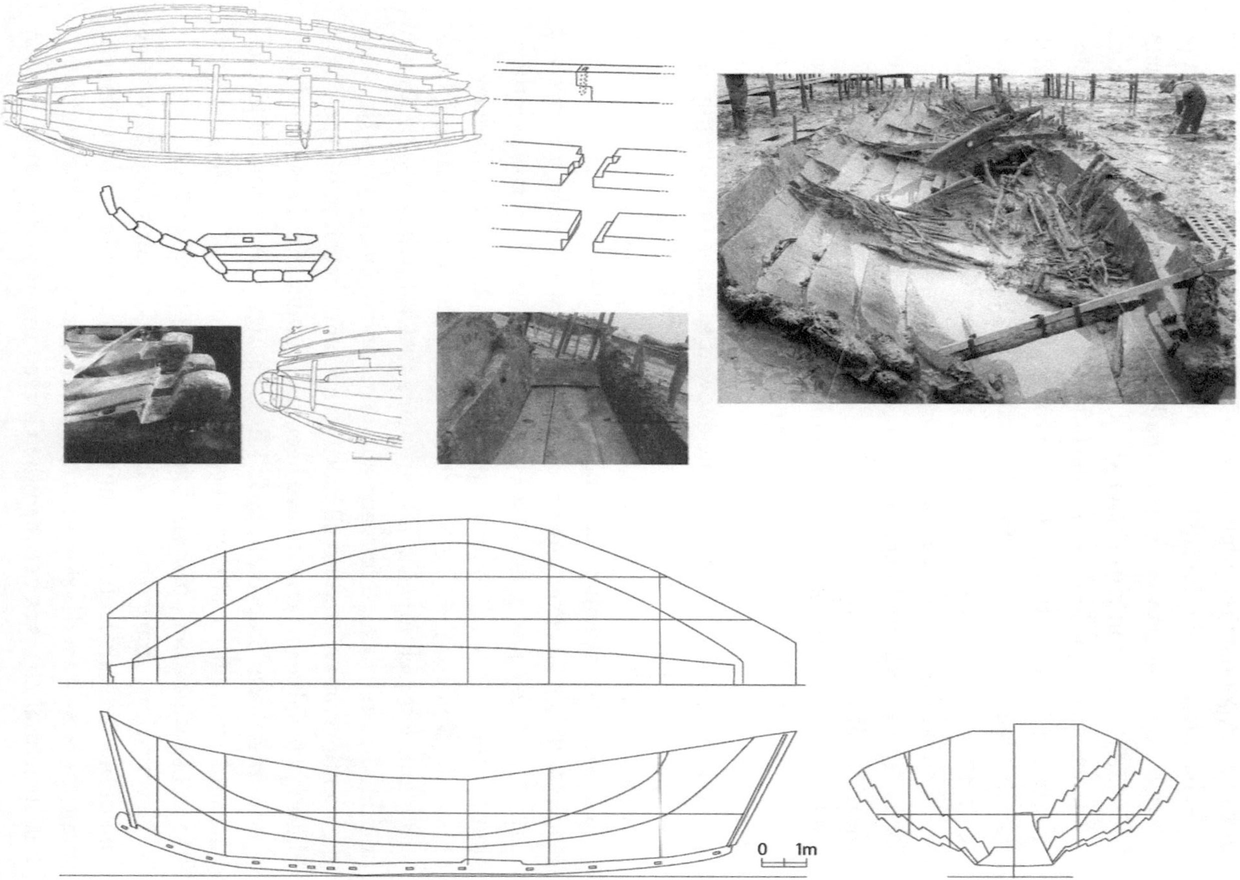

Figure 2.4. Anjwa ship, plan of the hull, cross section, schematic diagrams showing the butts of the bottom and side strakes, and photographs and an accompanying diagram showing bow and stern structures. (National Maritime Museum of Korea 2006; ship lines drawn by author)

of hull planks. Evidence of using putty for caulking and sealing the hull planking has been identified in this ship.

Jeremy Green (1986) has highlighted the distinctive construction methods of the Goryeo dynasty, compared with the features of twenty early Korean fishing boats reported by other researchers. Horace Underwood (1979) has identified hull planking of early twentieth-century coastal ships that relies on bent skewed wooden nails (known as treenails or trenails [figure 2.5]). According to Underwood (1979: 25), the holes are chiseled for the wooden nails in those modern fishing boats. In a similar way, the wooden tenons identified in the Goryeo ships must have passed through holes that had previously been cut. Further details of the driving method are described as follows: "The trenails are driven diagonally down from the outside of the upper plank and project on the inside of the lower planks. These ends are later sawed off, and the outer end counter sunk" (Underwood 1979: 25).

As Underwood notes, trenails are used almost exclusively in Korean boats even today, and these are almost invariably of oak, as are the anchor chocks, the pegs for the sweeps, and other fittings where a hardwood is required. The wood is not usually allowed to season thoroughly, as it thereby becomes too stiff to bend easily (Underwood 1979: 24).

Regarding the type of wood used for other components of the hull, Underwood has explained that Korean shipwrights preferred pine. A Korean researcher has examined the types of the wood used for different parts of Korean ships (Choi 2006: 28); details of wood species identification for the excavated ships are presented later this book. Underwood (1979: 25–27) describes the hull construction methods for fishing boats as a plank-fast technique, and he explains structural details with Korean shipbuilding terms used for each hull component. Questions arise about the identification of traditional naval terms (and their English spellings), although naval terms could have been used under the shipwrights' own unique practices: "Ship-building in Korea has never been reduced to a science and apparently has always been done by rules of thumb rather than to exact specifications, plans or lines" (Underwood 1979: 22).

Nevertheless, archaeologically identified ship remains dating from the end of the eleventh century to the end of the fourteenth century demonstrate the consistent development of Korean coastal ships. There are obvious structural changes between the Goryeo dynasty ships and modern

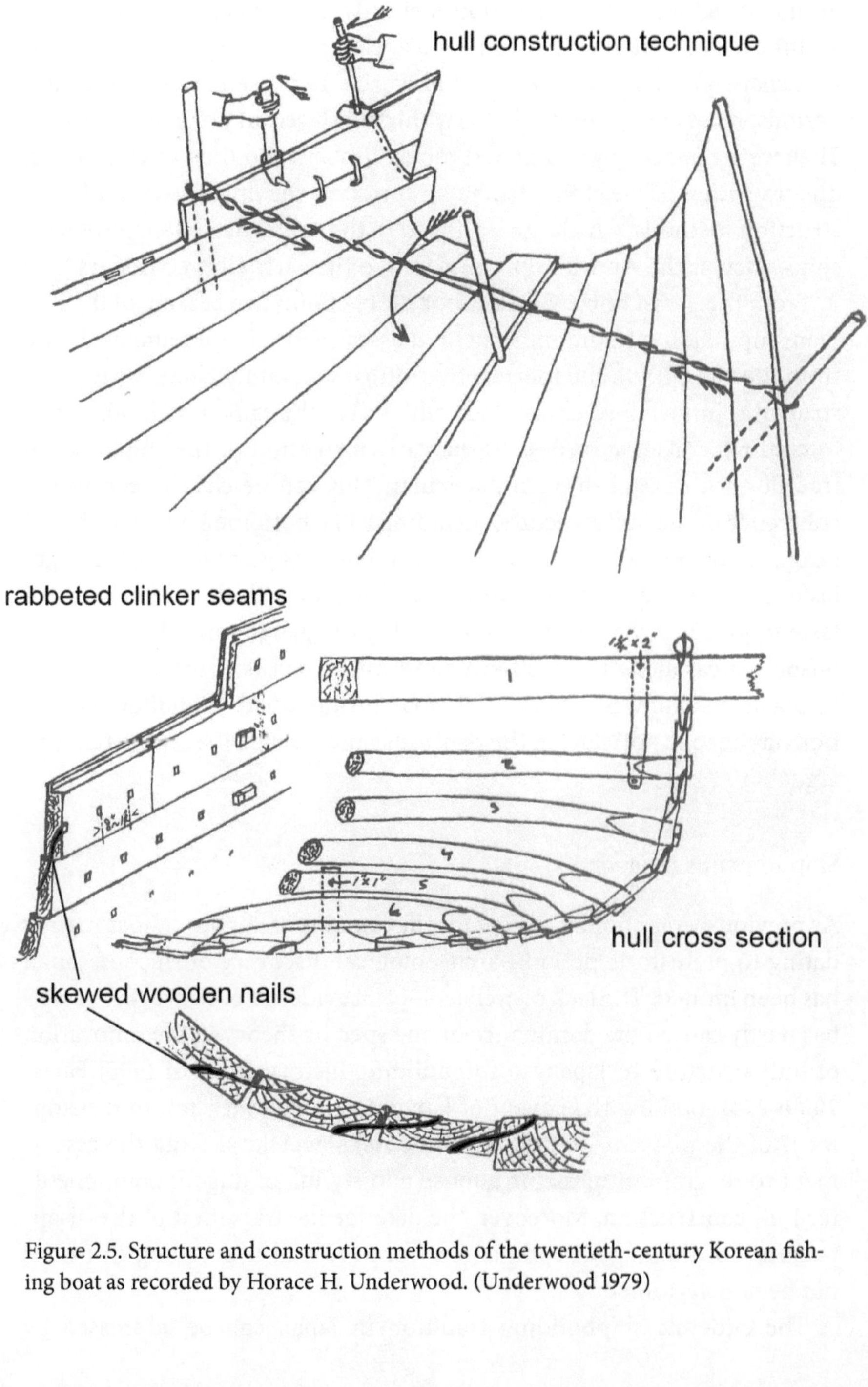

Figure 2.5. Structure and construction methods of the twentieth-century Korean fishing boat as recorded by Horace H. Underwood. (Underwood 1979)

Korean ships. L-shaped chine strakes represent the historical transition of hull structure and construction methods. The strakes were originally an innovative structure, but to produce chine strakes having such complex shape probably required a lot of work. Their use ceased in the later periods, most likely around the early thirteenth century or slightly earlier. They were replaced by chamfered robust timbers and then thick timbers. The transition of the chine structure represents the improvement of construction methods, which can be linked to the sophisticated shape of later ships, such as the Anjwa ship, compared to the early Goryeo traders.

From this point of view, we cannot simply point to a ceasing of the Korean shipbuilding tradition. Structural assessment of hull remains dating from the eleventh to the fourteenth centuries certainly allows us to trace structural innovations chronologically, and at the same time it allows us to confirm conservativeness in the early formation of the shipbuilding tradition for coastal ships in particular. This can be clearly seen in the coherence of the hull structure, including a flat-bottomed hull and thwart beams, and consistency in construction methods is evident in the edge-fastened planking used to form the clinker hull with the use of wooden fastenings. Discovery of the Goryeo ships is geographically limited to offshore areas along the southern and western coasts of the Korean Peninsula in the Yellow Sea. How influential the identified shipbuilding tradition has been historically in the peninsula and outside the region remains arguable.

Ship Remains in Japan

As previously mentioned, except for the remains of dugout canoes mainly dating to prehistoric periods, archaeological discovery of ships in Japan has been limited. The lack of archaeological evidence of seaworthy vessels has partly caused the dominance of one specific theory in the innovation of hull structure in Japanese shipbuilding history (Adachi 1998; Farris 2009: 274). Linking the advent of Japanese native watercraft to development of the planked-up dugout canoe risks oversimplifying the assessment to diagramming use or nonuse and styling of dugout components for hull construction. Moreover, the date for the transition of the disappearance of the dugout component from the hull of seagoing ships has not been determined.

The endemic shipbuilding tradition in Japan can be addressed by reviewing one component of the hull of Japanese coastal traders dating to the postmedieval Edo period (AD 1603–1868). Two large rudder posts, probably associated with these coastal ships, have been identified; they are the only available archaeological resources we have for assessing shipbuilding traditions for seaworthy ships in Japan. They have been independently identified, and no hull remains are associated with them. They were buried in the sediments of adjacent historical harbors. There are no relevant archaeological survey records to examine for associated context, such as the existence of shipyards, or for clues to assess whether the rudders were deliberately abandoned or became wreckage accidentally. In terms of a thematic study of rudders, their analytical utility is not enough for one to deduce whether their structural innovations followed those of the previous periods. In general, depending on a large rudder post or rudder stock with a tiller is perceivable in the tradition of East Asian shipbuilding technology. Diffusion and interaction are presumed in the development of an axial rudder within the region. However, an axial rudder in Japanese coastal ships is a hanging rudder fixed by ropes in the stern, as distinct from a transom-mounted rudder slotted into the transom (as is evident on some Chinese ships).

Rudder Post from Bonotsu Harbor

According to an online database of underwater sites in Japan produced by the Asian Research Institute of Underwater Archaeology, a large wooden rudder stock was discovered in the riverbed of the Kushi Imamura River in the town of Bonotsu. It is now exhibited in the Bonotsu Historical Resource Center in the city of Minamisatsuma. Bonotsu is a historically significant harbor developed before the tenth century that continued to be used until the Edo period for both domestic and international trading. Analysis of the data permits a brief definition of the features and origin of the rudder post. Its remaining part measures 8.2 meters in length, and the largest cross section is 490 × 230 millimeters, although both ends are damaged. Chemical preservation has been conducted on the timber. The wood has been identified as *Quercus acuta*, which grows widely in East Asia. The rudder is presumed to have originated from a coastal ship called a *sekibune*, but this does not seem to have been ascertained with concrete evidence. A sekibune is generally classified as a midsized coastal ship developed during the Japanese medieval period as a battleship and on some occasions was used in postmedieval periods to carry members

Figure 2.6. Scale model of a *sekibune* that was made based on historical and iconographic resources. (Courtesy of Matsuura Historical Museum; photo by author)

of the peerage (figure 2.6). The details of this rudder stock, including how it was dated and classified as from a sekibune, are not clear. The configuration of the sekibune can be vaguely depicted based on iconographic resources and historical texts. In general, it is presumed to reflect Japanese representative construction methods using multiple bottom planks rather than a keel, with thwart beams penetrating the hull planking and fastened by iron nails and brackets. One type of sekibune constructed as a battleship appears to have had bulwarks of substantial size and a castle-like superstructure running from the bow to the stern, and its multiple oars enhanced not only propulsion but also maneuverability.

Rudder Post from a Japanese Coastal Ship

Another example of a rudder post from a Japanese ship comes from Okitsu town in Chiba prefecture. The large timber is believed to have been a part of the rudder from a traditional Japanese coastal ship used during the Edo period. The timber was discovered in the estuary of a small river flowing into Okitsu Beach during the development of an extension of the harbor (figure 2.7). Historically Okitsu served as an anchorage for traders

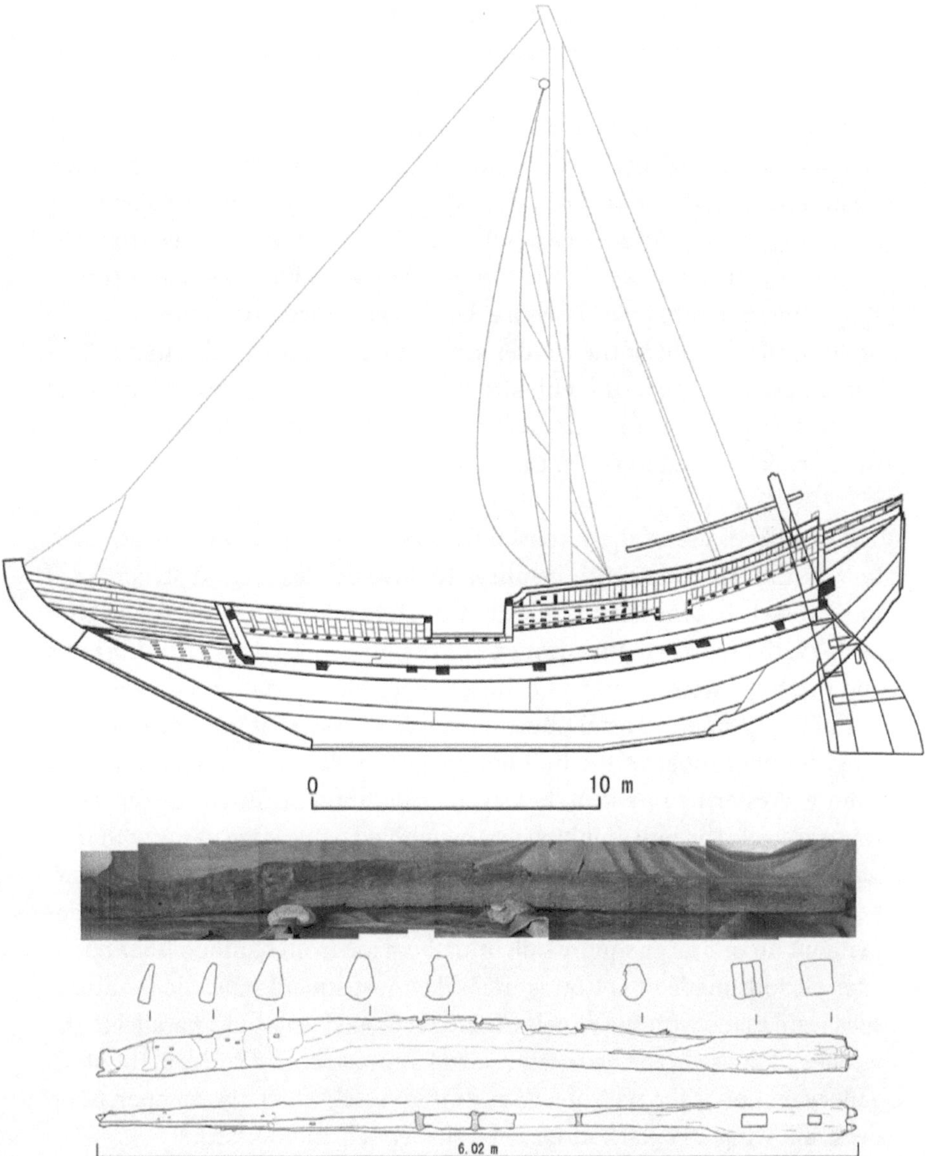

Figure 2.7. Drawing of the early nineteenth-century *bezaisen* and drawing and photograph of the remains of the rudder post. (Produced by author)

sailing along the east coast facing toward the Pacific Ocean. The historical background of the place of discovery as well as the configuration of the rudder post indicate that it was used for a small coastal trader. The rudder post is 6.02 meters in length and is made of hardwood. When discovered it was in a well-preserved condition, but lack of proper conservation treatment caused it to deteriorate and split, with cracking and shrinking appearing on the surface. Two holes in the middle are damage from the harbor construction work. The top part of the rudder post has a square shape, and a square hole allows a tiller to be inserted. From the top to the bottom, the section of the rudder gradually changes to a flat shape. This part could have had a rudder blade attached, though it is now missing. In the bottom part of the rudder, a few corroded iron pins can be seen; these were probably used to fasten the rudder blade to the rudder post.

The heyday of use of these coastal traders was the mid-seventeenth century to the late nineteenth century. The overall design and structure of these vessels became standardized through the period. By the nineteenth century, the representative features of the hull showed more uniformity, and the ships were known as *bezaisen*. Regarding structural characteristics of the Japanese coastal traders, it has been said that the spine (a main longitudinal timber of the hull bottom) of the bezaisen is distinguished from a Western ship's keel; however, structurally it functions the same way as a keel. The hull planking was assembled using edge joints with iron nails. The transverse strength of the hull was secured by several thwart beams distributed along the length of the hull. Propulsion was highly dependent on one large square sail, initially made from bamboo fiber but in later periods made of cotton fabric. Like most square sails, the bezaisen's sails were made with the panels of cloth arranged vertically, parallel to the leeches of the sail. The bezaisen used a large suspended rudder. The stern gallery section is the typical feature of this vessel, which has an open poop deck, allowing an axial rudder to be suspended and hoisted.

As mentioned earlier, there is an indication in a Chinese historical text that shipbuilding in Japan could be characterized by the construction of flat-bottomed ships. In terms of hull components, the main transverse structure in Japanese ships was thwart beams. Looking at another representative hull component of East Asian seagoing ships, one can see that bulkheads were never adopted in either Japanese coastal ships or Korean traders.

A question arises about similarities in hull components identified in the two regions, such as the use of beams as the main transverse components. The possibility of Japanese shipbuilding having adopted some technologies from outside has not been seriously addressed in the existing theories. A few people outside Japan have alluded briefly to this possibility, for instance, in mentioning that "Japanese ships seem to follow a similar style to Korean ships" (Green 1986: 4). Underwood also remarked that many shipwrights were sent from Silla to Japan before the tenth century to transfer shipbuilding skills (Underwood 1979: 23). Verification of early historical interactions between the two countries that affected the development of the Japanese shipbuilding is expected to occur through archaeological discovery. It will be relevant in discussing whether a linear development theory is valid to explain the advent of Japanese shipbuilding traditions. The characteristics of the hull structure of Japanese coastal traders can more reasonably be understood by defining them as flat-bottomed ships exhibiting thwart beams, which synergistically form a sphere of the shipbuilding tradition inclusive of the Korean Peninsula.

3

A Lasting Tradition in Northern China

In China more than twenty ships' hulls have been excavated on land and underwater (Kimura 2010a). This chapter presents some of these ship remains chronologically according to their dynastic periods, from the Sui dynasty (AD 589–618) to the Song dynasty (AD 960–1279). Many of them have been dated based on associated artifacts, which may not be sufficient to provide an exact date for the hull. Thus, no absolute building dates for the ships have been established.

The data set of hull remains indicates a coherent tradition of practices in building flat-bottomed ships in northern China. We can also recognize that several excavated hulls dating to the early period are those of riverine ships. They illustrate some representative structures and construction methods that had been widely adopted by then, such as bulkheads and the use of iron nails for fastening. The archaeological data are useful for comparatively addressing the endemic features of early shipbuilding in northern China as distinct from central and southern China. Historical and political background about the dynasties provides insights regarding factors that might have affected shipbuilding in early periods. In addition, iconographic and historical resources provide complementary explanations that aid in in identifying a sphere of lasting shipbuilding tradition from north of Hangzhou Bay around the Yellow Sea.

Archaeological Evidence of Early Chinese Ships

One of the earliest examples of iron used for ship construction is among the ship remains discovered in the Sui dynasty tomb of Zhongshanwang (中山王), in Pingshan, Hebei Province (Xi et al. 2004). The tomb has a funeral chamber inside, in which was found a well-constructed ship in use as a coffin. The ship is 13.1 meters in length and 2.3 meters in width, and its depth is 0.8 meters (figure 3.1). Its basic structure consists of several planks

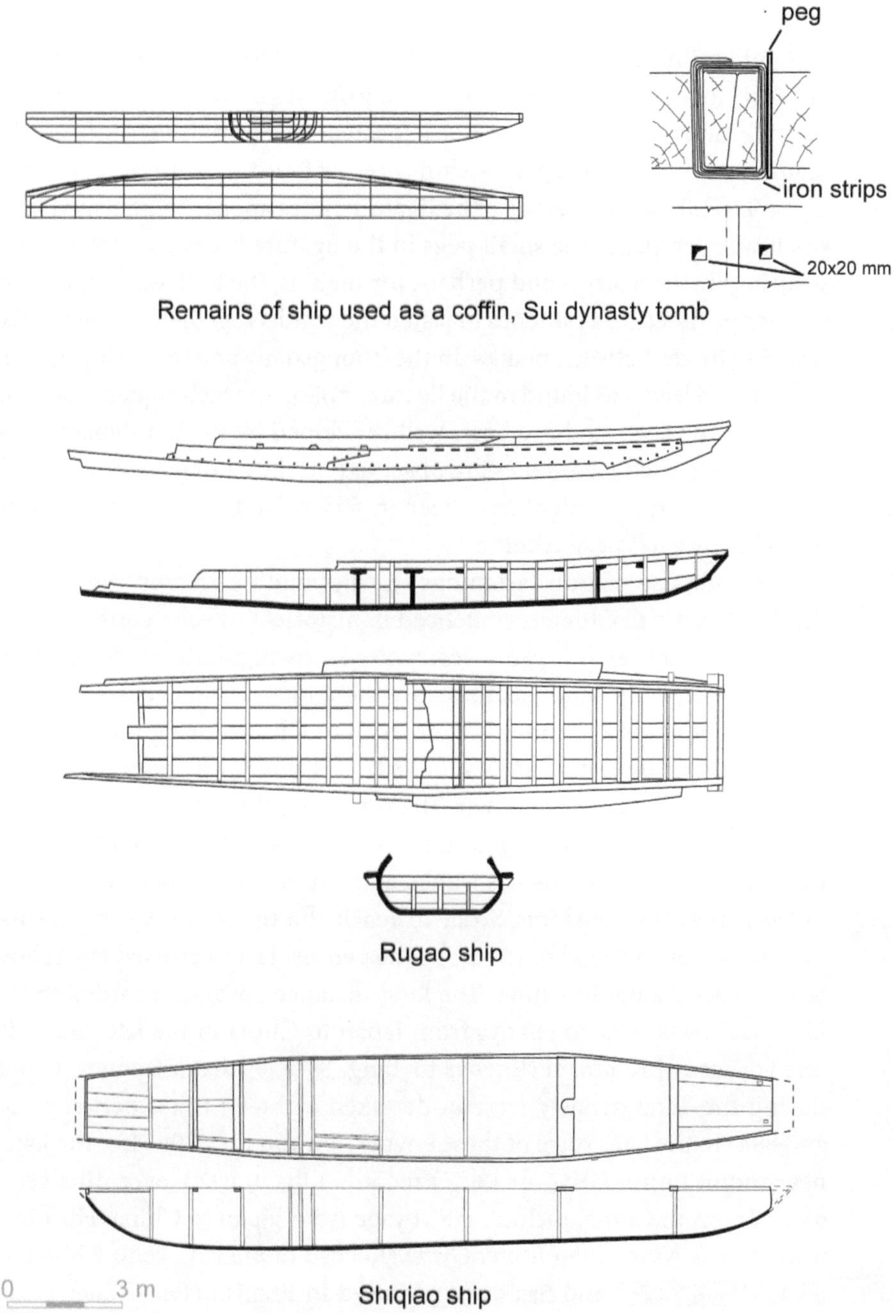

Figure 3.1. Excavated Sui and Tang period ships. *From top*: Reconstructed ship lines of the ship used as a coffin, discovered in the Sui dynasty tomb of Zhongshanwang, and metal fastenings for its planking; Rugao ship plan and cross section; and Shiqiao ship plan and cross section. (After, from top: Xi 2000; Wang 2000; Xi 2000)

400–600 millimeters wide and 100–150 millimeters thick. The planks are fastened together with iron strips. The iron strips are threaded through 20-millimeter-square ligature holes located 40–50 millimeters from the seam to tie the planks together with three or four turns. These iron strips are at 0.8–1.0 meter intervals. The ligature holes appear to be sealed with small wooden pegs. The small pegs in the ligature holes are probably for securing the iron strips and perhaps for making the hole watertight. The iron strips function as stitches to fasten the planks together and also resist shearing forces between planks. In the Zhongshanwang tomb ship, a sealing of fused lead was found in the ligature holes. The techniques of sealing and caulking seem to have been well developed by the Sui dynasty. The fact that two such techniques are observed on the Zhongshanwang tomb ship indicates that it might have been in service on the water for a certain period before its use as a coffin.

The wide use of iron fastenings in shipbuilding around the time of the Sui dynasty in China is evidenced in historical travel records. During the seventh and eighth centuries, many ships regularly made sea voyages between China and Japan. Before then, interaction occurred by delegating diplomats and monks from Japan to China during the Southern and Northern dynasties (AD 420–589) and the Sui dynasty. Details of the ships involved are unknown. The route that the Japanese representatives took did not entail long-distance voyages, and ships departing from what is now Osaka (known as Nanba at the time) got to Kyushu and then sailed across the Tsushima Strait to reach the tip of Korean Peninsula. The ships then voyaged north along the peninsula and crossed the Yellow Sea to reach Yantai in China. The long-distance voyages recorded in the historical texts refer to envoys from Japan to China in the later periods. The voyages of Kentoshi (Envoys to Tang, 遣唐使) from Japan to China during the Tang dynasty were undertaken eighteen times between 630 and 894. In the later stage of these envoys, around the 830s–40s, the Japanese monk Ennin (円仁 or 圓仁) recorded his journey over nine years as a pilgrim to China, including a voyage from Japan to China. His diary is known as *Nitto Guho Junrei Gyoki* (Record of pilgrimage to China, 入唐求法巡礼行記) and has been reviewed in English (Reischauer 1955). According to the record, Ennin's envoy ships confronted difficulties, and his diary depicts his hazardous voyage. The problems may have resulted partly from the paucity of navigation techniques and partly from embarking on the journey in such a fragile ship—he noted that the "iron strips came off by wave smashing (平鐵（鎹？）為波所衝悉脱落)." The "iron strips" to which he referred may mean strips like those used on the Zhongshanwang tomb ship, in which case the manuscript may indicate that the planking of the ship in which Ennin traveled mainly used iron fastenings, though an argument on detailed configuration of the fastenings is speculative.

Tang Shipbuilding in the Riverine Tradition

By the Tang dynasty period in China, shipbuilding technologies preceding the advancement of seafaring had been accumulated in the tradition of building ships for riverine use, as we can see by looking at the remains of river ships dated to this period. They have been studied only cursorily; the following is a summary of the limited information available.

Rugao Ship (如皋)

According to Chinese sources (Wang 2000; Xi 2000; Xi et al. 2004), the Rugao ship was found at Puxi in Ruguo city, Jiangsu Province, during reclamation in 1973 (see figure 3.1). The well-preserved hull measures 17.3 meters in length and 2.6 meters in width. The remaining part consists of bottom planks, bulkheads, and some strakes from the sides of the hull. The bottom is a flat shape suitable for river transportation. Its blunt bow and aft parts are slightly narrower than the midship. The Chinese researchers' reports (cited above) indicate that the basic joinery of the bottom timbers appears to be mortise-and-tenon style, but the details are unclear. Mortises and tenons are also said to be used for the bulkheads and hull planking.

The hull profile shows that bulkheads divide the hull into eleven holds. The fifth, sixth, and seventh bulkheads are very close together and create two very small holds. The breadth beam of the vessel is approximately 2.8 meters and measures approximately 0.95 meters athwartships in the bow and stern. Chinese researchers have mentioned that there are doorways through the sixth to the eighth holds, yet these bulkheads produce watertight compartments. The Chinese researchers' reports state that these holds could have been used as a living space covered by bamboo sheets in the absence of decks. The hull planking is clinker-built. The thickness of the planks is 40–70 millimeters, and the bottom planks are 80–120 millimeters thick. The length of iron nails is 165 millimeters, and the nailhead measures 15 millimeters in cross section. Iron nails are used for the fastening of the hull planking, with a pattern of nails driven at 60-millimeter intervals. The seams are sealed by putty. Associated artifacts, mainly ceramics from the Yue kiln, allow the Rugao ship to be dated to the Tang dynasty. Chinese researchers have concluded that the ship was used around the reign of Gaozong (AD 649–83), the third emperor of the dynasty.

Shiqiao Ship (施桥)

During a river development project in 1960 at Shiqiao in Yangchan, Jiangsu Province, one wooden ship was found with a few dugout canoes (see figure 3.1). According to Longfei Xi (2000), the ship shows a sturdy structure. Although its stern section was missing, the remaining part of the hull measures 18.4 meters in length, 2.4 meters in width, and 1.3 meters in depth. The transverse structure consists of a combination of beams and bulkheads (the exact number of holds divided by the bulkheads is unclear). Mortises and tenons and iron nails appear to have been used for the hull planking. The nails are 170 millimeters in length, with a head measuring 20 millimeters across, and they are driven into hull planks at intervals of approximately 250 millimeters. The nails are used for the construction of the bottom as well as the sides. The exact date of the ship is disputable. Initially it was dated to the Song dynasty based on artifact analysis, but later the majority of researchers reassessed this, concluding that it dates back to the Tang dynasty. The depth of the ship is not great, and the structure indicates that it was originally used for river transportation.

Technology of Riverine Ships

All the identified remains of Chinese ships from before the tenth century seem to involve ships for riverine use, not specifically designed for coastal or sea voyages. They are dated to the late periods of the Tang dynasty. The state of development of riverine infrastructure during the Tang dynasty continued from the previous dynastic period, and presumably it propelled Chinese shipbuilding industries to construct inland watercraft for transportation in rivers as well as canals.

These river craft exhibit techniques of ship construction that developed before the tenth century. Fastening methods using iron had become standard by the Tang dynasty period. Compared with other metals, iron has been predominant in China for a long time. The use of putty for sealing and luting likely became widespread during the Tang dynasty; use of some kind of putty probably dates back to earlier periods. It was also combined with caulking materials in the form of vegetable fiber pushed into the seam after construction, the putty being used to seal over the caulking.

With the development of planked hulls, the transverse structure of the ship also evolved. As noted, the use of bulkheads in constructing riverine ships is evidenced in the archaeological discoveries. Bulkhead structure and construction methods seem to have been well established in riverine vessels by the ninth century, before being employed in seagoing ships. In China, unlike other areas of East Asia, these components had already been developed by the time the Chinese commenced seagoing. Little information is available on what types of ships were used in the early stage of seafaring. Whether riverine ships were used or there were changes in ship structure and shape before seagoing commenced is still an open question. Internal conflicts and political disorder occurred in China from the end of the Tang dynasty to the Five Dynasties and Ten Kingdoms period (AD 907–60), and historical records contribute little to any discussion of the type of ships used during these periods. Presumably, seagoing Chinese ships developed in a relatively short period of time. Diffusion of the use of bulkheads in shipbuilding was limited to the northern riverine area and coast of China. Each nation within East Asia started to show separate shipbuilding traditions by the tenth century, as reviewed in the previous chapter.

After the fall of the Tang dynasty, China experienced a series of wars that brought stagnation in social activities and could have negatively affected seaborne trade. Less activity in constructing coastal and seagoing craft may have led to stagnation or ceasing of technological innovations, in turn implying less stimulation from the low volume of seaborne trade. Chinese merchants made greater inroads in later periods. Raising their presence seems to have occurred within a few hundred years or even less, presumably around the period of the Northern Song dynasty (AD 960–1127). Archaeological discoveries of Song dynasty ships are more numerous.

Early Song Shipbuilding in Riverine and Coastal Tradition

Dazhi Boat (大治舟)

According to a short archaeological report (Ji 1987), ship remains were discovered in 1978 during the dredging of the Dazhi River, Nanhui district, in Shanghai city. The Cultural Properties Administration Board in Shanghai city conducted the excavation at the site (figure 3.2). Although the upper part of the hull is missing, the bottom was well preserved when discovered. The remaining part measures 16.2 meters in length and 3.9 meters in width. The cross section of the hull bottom is flat. The bottom consists of seven strakes. The strakes comprise three planks, joined by lap joints. Three internal runs of planks double the bottom planking, though the remains are partial.

Eight bulkheads divide the hull into nine holds. The spacing of the bulkheads varies from 0.8 to 2.9 meters (2.9 m, 1.3 m, 1.9 m, 1.7 m, 1.8 m, 1.4 m, 0.8 m, 1.3 m, and 1.6 m). There is no report of limber holes. Iron nails were used to fasten the hull planking to the bulkheads, and putty made from limestone and oil was used to seal the iron nailheads. A notable feature is a hole with a diameter of 200 millimeters through the median plank in the bow (if the bow is correctly identified on the basis of the position of a mast step). This is said to be what is called a *mao chayan* (hole to drive anchor through, 錨挿眼 [Ji 1987: 177]). This type of anchor was reported as a "stick-in-the-mud" anchor in Worcester's (1971: 89) study; it is a pole used in Chinese river boats as an anchor by being driven through the bottom of the hull into the bed of the river. There is one mast step, and it is attached to the third bulkhead, which has one recess to hold a mast. Discovered in a hole located at the first hold were twenty-four copper coins inscribed Taibin Tongbao (太平通寶), which represents the reign of the second emperor of the Northern Song dynasty. In the hole a silver hairpin was also found. On the basis of the coins and one glazed ceramic bowl, the ship has been dated to the Song dynasty. It was probably used for short-distance coastal transportation before, for some reason, it was abandoned (Ji 1987: 177).

Yuanmengkou Ship (元蒙口船)

According to Longfei Xi (2000: 152–57; 2008: 66–67), Song dynasty ship remains were discovered in 1978 at Yuanmengkou, at Dongtantou in the Jinghai district in Tianjing Province. The remains of the ship are 14.6

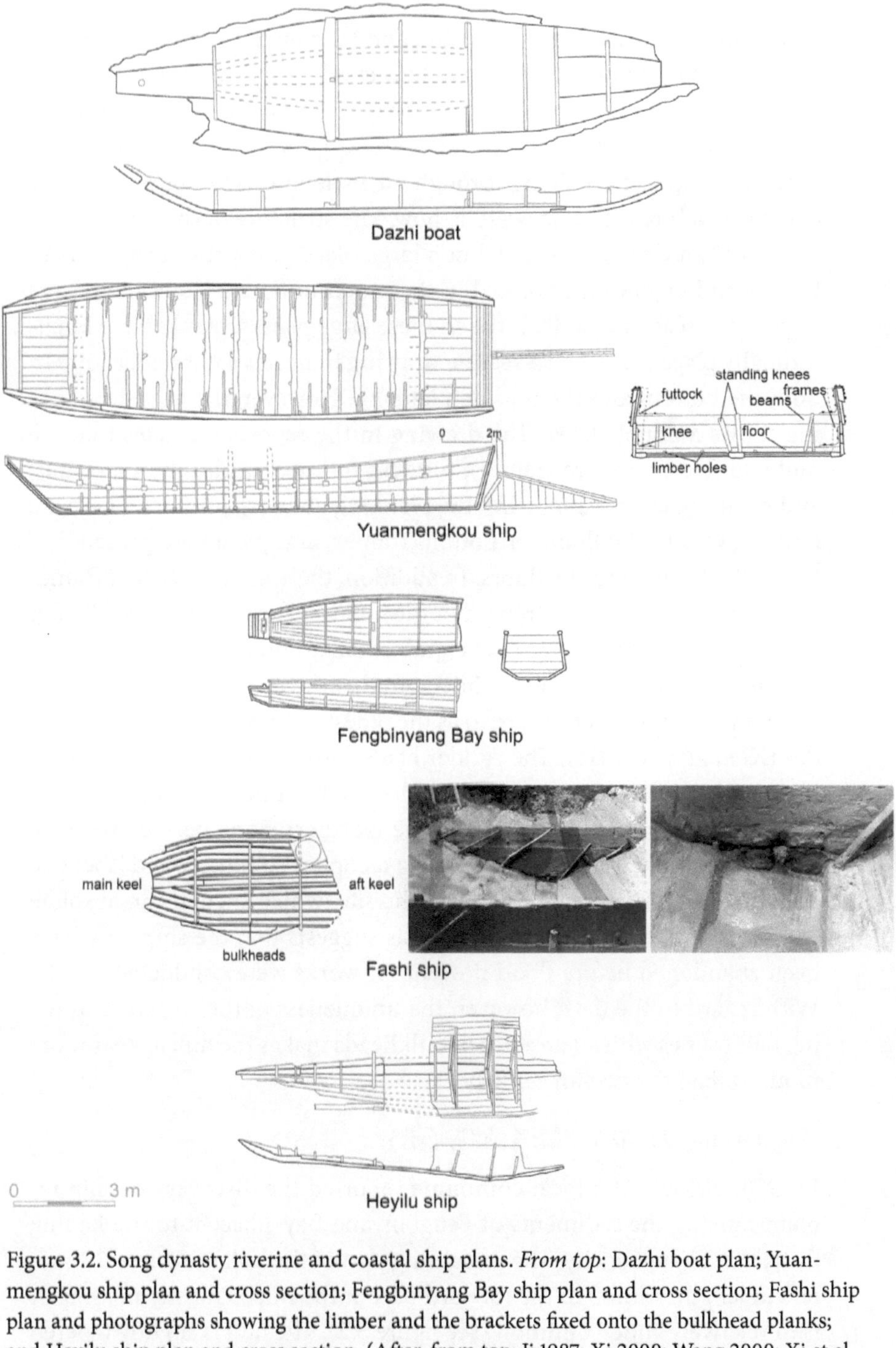

Figure 3.2. Song dynasty riverine and coastal ship plans. *From top*: Dazhi boat plan; Yuanmengkou ship plan and cross section; Fengbinyang Bay ship plan and cross section; Fashi ship plan and photographs showing the limber and the brackets fixed onto the bulkhead planks; and Heyilu ship plan and cross section. (After, from top: Ji 1987; Xi 2000; Wang 2000; Xi et al. 2004 and courtesy of the Western Australian Museum; Gong et al. 2008)

meters in length, 4.1 meters in width, and 1.2 meters in depth. Except for the upper part of its port side, the hull is relatively well preserved. The displacement of the original ship has been estimated to be approximately 38 tons (Xi et al. 2004: 112). The shipwreck has a flat bottom and shows a rectangular section shape throughout its length (see figure 3.2). It is effectively a box with a scow-style bow. The structure of the bottom hull has not been clearly reported, but a large piece of flat timber appears to have been longitudinally placed in the center of the hull, and there were large chine planks as well. Between these large planks, planks were longitudinally placed. The cross-section drawing indicates a flush edge joint in both the bottom and the topside planking. However, the topside planking shows a double layer. The drawing in the reports indicates that the hull had floors (bottom frames), futtocks or knees at the chines, beams, and standing knees supporting the planking above the beams. There are limber holes in the floors or bottom frames, and pillars are present between the beams and the floors. In addition, there are twenty half frames between the beams. The strength of the hull was further enhanced by top timbers on the fifth beam to the eighth beam around the midship.

Chinese researchers have pointed out that the most significant discovery is a wooden rudder, regarded as the oldest example of a balanced rudder (Xi et al. 2004: 114). The rudder blade forms a scalene triangle measuring 3.9 meters along the bottom. The height of the rudder post is 2.19 meters. The ship could be dated to the Northern Song period. The date was initially determined by identifying a copper coin inscribed Zhenghe Tongbo (政和通寶) recovered from the site, which suggests an absolute date of 1111, and sedimentation analysis suggests that the ship may have been abandoned before flood prevention works were conducted in 1117. With regard to the date, however, the uniqueness of the structure in using half frames with an absence of bulkheads makes the hull appear more modern than the twelfth century.

Fengbinyang Bay Ship (封滨杨湾沉船)

In 1978, officials of a local commune reported the discovery of ship remains among the sediments of Fengbinyang Bay adjacent to the Jiading district in Shanghai city. The remaining part of the hull measures 6.2 meters in length. Although the forward part of the ship is mostly missing, it is in relatively good condition (see figure 3.2). The hull is slightly tapered from the midship toward the bow, yet it forms a square transom bow that has a platform at the deck level. There are seven bulkheads constituting eight compartments, and a mast step is placed on the forward side of the fourth bulkhead. Iron brackets, known as *guaju*, are used to fasten hull planking to the bulkheads (Wang 2000: 120–21). Some butts of hull planks show a half-lap joint. The cross-section shape shows a flat bottom made of two strakes, double chines, and some tumble home of the topsides above the second chine. One of the notable features of the ship is two longitudinals beneath the bottom planks. The two longitudinals, semicircular in cross section, run along the hull as small bilge keels or bilge runners. It has been pointed out that these are likely to provide longitudinal strength to the hull, and they appear to be small runners designed to protect the bottom planking during launching and beaching (Wang 2000: 120–21). Found inside the hull were bricks and yellow-glazed ceramics, probably from the Jizhou kiln in the Jiangxi region. These artifacts were used to determine the date of the ship as late thirteenth century. An iron pan and sword also indicate a link to this period.

Fashi Ship (法石船)

A shipwreck was found near Fashi in the vicinity of Quanzhou Bay in 1982. A portion of the hull was located under a building, and only a partial excavation was implemented (Green 1997a: 19; Xi et al. 2004: 124–25). An exposed part of the hull was reburied for in situ preservation. A midship portion toward the stern was revealed during the excavation. The hull had a keel, and three bulkheads were also identified. The exposed keel consists of part of the main (midship) keel and the aft keel. Bilgewater could run along the bilge through limbers in the bulkheads. The Fashi ship shows evidence of the use of wooden brackets (or pegs) to fasten bulkheads to hull planking. One wooden bracket measures 720 millimeters in length, and its cross section is 60 × 60 millimeters at the hull planking and tapers to 20 × 30 millimeters at the inboard end. The brackets are driven from the outer surfaces of some of the hull planks and attached to the forward sides of the bulkheads. This fastening method, similar to that noted in the Fengbinyang Bay ship, is also observed on the later Chinese seagoing trader known as the Quanzhou ship (see chapter 4) and is more or less identical to the use of wooden brackets in the Shinan shipwreck (see chapter 5). Iron nails are used to fasten the bulkhead planks. A portion of the recovered hull including a bulkhead plank is under chemical treatment in the Quanzhou Ancient Ship Gallery of the Museum of Overseas Communication History (figure 5.7).

Heyilu Ship (和义路船)

In 2003 the remains of a watercraft were found during a rescue excavation conducted by the Cultural Relic Preservation Administration Center of Ningbo and Archaeological Research Institute of Ningbo at the south side of the site of Heyimen Wengcheng, or Turret of Heyi gate (Gong et al. 2008). Although the stern of the hull is missing, the remaining part is relatively well preserved and measures 9.2 meters in length and 2.8 meters in width (see figure 3.2). The bow is fairly sharp, and the cross section of the lower hull is a V shape. The forward keel and main keel remain: the midship cross section of the main keel is rectangular and measures 300 × 100 millimeters, but toward the bow the keel cross section changes to a triangular shape to fit to the forward keel. Nine bulkheads remain, and sturdy frames are attached to them. The bulkheads appear to be irregularly distributed. Each bulkhead consists of three to four planks fastened by iron clamps. There are limber holes in the bottom planks of the bulkheads. Seven strakes remain on the port and starboard sides in the midsection of the hull. Hull planks measuring 400–600 millimeters are edge-joined by skewed iron nails. A mast step has not been found, and a drawing of the reconstructed ship in the report suggests that its propulsion relied on a *yuloh* (Gong et al. 2008: 187). On the basis of ceramics recovered from the inside of the hull, the ship has been dated to the Southern Song dynasty (AD 1127–1279), and it could have been used for short-distance transportation in the harbor or along coastal areas.

Technological Development from the Riverine to the Coastal and Harbor Ships

As the preceding descriptions demonstrate, several ships dating back to the Song dynasty have been excavated on land, and they were used as both riverine and coastal ships. Short archaeological reports published in Chinese are the primary resources for detail about these ship remains. Some seem to show ship types that could have been widely used, not only in riverine settings but also in estuarine areas or for lighterage in ports; these include the Dazhi boat, the Yuanmengkou ship, and the Fengbinyang Bay shipwreck. They were discovered in the late 1970s when only limited preservation could be undertaken. The locations of these ship remains are around Shanghai on the estuary of the Yangtze River flowing to Hangzhou Bay and Tianjing in the northern part of China. In cross section these ships present a flat or flat to round bottom, corresponding with the generic idea that northern ships were used either inland or in the open waters typically observed north of Hangzhou Bay. Frames and bulkheads are used as transverse components. The use of the axial rudder and mast steps is suggested to date back to earlier than the Song dynasty.

Besides fastening relying on iron nails, the use of brackets to secure hull planking to bulkheads, which is seen on the Fengbinyang boat, is noted. Coherent use of iron fastenings and the emergence of various types of iron fastenings characterize the Chinese shipbuilding tradition. Iron fastenings include nails, brackets, and clamps, and each type can be further subclassified into different shapes. These have been reviewed by Chinese scholars (Xu 1985; Xi et al. 2004: 123–25; Zhang et al. 2004: 288–91). In these studies, iron brackets have been defined as a more advanced form of technology evolved from wooden brackets. The Fengbinyang Bay shipwreck is the earliest archaeological evidence of the use of metal brackets. As there is no evidence of wooden brackets predating Fengbinyang, the evolution mentioned is speculative. Continuing use of wooden brackets can be observed over centuries. Use in the fastening of hull planking to bulkheads is the same for wooden brackets and iron brackets. The Fashi ship discovered near Quanzhou city in Fujian Province in the 1980s uses brackets made from wood. The cross section of its hull shows that the bottom of the ship is not flat (a flat bottom is regarded as a feature suitable for use in shallow and tidal waters). The Heyilu ship excavated in Ningbo city in 2003 was probably used for inshore transportation and for lighterage in ports. It has a sharp bottom with considerable deadrise. The Heyilu ship indicates the advent of diverse hull types: not only flat-bottomed but also V-bottomed ships were used on inland and inshore waters by the late Song dynasty.

The Flat Bottom in the Han Chinese Shipbuilding Tradition

The configuration of riverine ships dating to the Song dynasty period can be seen in the picture scroll by Zhang Zeduan (AD 1085–1145) titled *Qing Ming Shang Ho Tu* (Along the river during the Qingming festival, 清明

上河圖). The scroll, widely recognized as an important resource depicting details of contemporary ships, has been reviewed in previous studies (Needham 1971; Green 1997b). It graphically depicts different types of river ships. Some of the features described in this scroll are useful for assessing the excavated ships, especially as regards rigs, sails, shipboard items, and the superstructure and stern structure of the ships depicted. The historical and archaeological resources appear to correspond with each other.

I propose that the utility of ships in river transportation in early China was one of the most critical factors propelling the formation of the regional shipbuilding tradition. The linkage of the flat-bottomed ship solely with topographical factors has been more emphasized in previous studies on the Chinese shipbuilding tradition (Worcester 1966: 16–17; McGrail 2001: 347–48; Van Tilburg 2007: 62). A well-known concept is that the very different waters of the northern and southern regions of the Chinese cultural sphere led to the development of two distinctive types of vessels along the Chinese coast. This has been mentioned with reference to Chinese historical records such as the *Chou Hai Tu Bian* (Illustrated seaboard strategy, 籌海図編), written in 1562, and the *Wu Bei Zh*i (Treatise of armament technology, 武備志), written in 1621. Joseph Needham (1971: 429) introduced the passage in the *Ri Zhi Lu* (Daily additions to knowledge, 日知録), originally written by Guyanwu (顧炎武) in 1672, as follows:

> The sea-going vessels of Chiang-nan are named "Sha-chuan (sand-ship)" for as their bottoms are flat and broad they can sail over shoals and moor near sandbanks, frequenting sandy (or muddy) creeks and havens without getting stuck. . . . Chekiang ships . . . (are built in the same way) and can also sail among sandbanks, but they avoid shallow waters as they are heavier than the sand-ships. But the sea-going vessels of Fukien and Kuangtung have round bottoms and high decks. At the base of their hulls there are large beams of wood in three sections called "dragon-spines" (lung ku). If (these ships) should encounter shallow sandy (water) the dragon-spine may get stuck in the sand, and if wind and tide are not favourable there may be danger in pulling it out. But in sailing to the South Seas (Nang-Yang) where there are many islands and rocks in the water, ships with dragon-spines can turn more easily to avoid them. (Needham

> 1971: 429, as translated from *Ri Zhi Lu*, vol. 29, annotated by Xie Zhanren)
> 江南海船．名曰沙船．以其船底平闊．沙面可行可泊．稍擱無凝．．．．浙江海船．．．亦能過沙．然不敢貼近淺處．以船身重於沙船故也． 惟閩・廣海船． 底圓面高． 下有大木三段．貼於船底．名曰龍骨．一遇淺沙．龍骨陷於沙中．風潮不順．便有疏虞．蓋其行走南洋．山礁叢雜．船有龍骨．則轉灣趨避．（日知録集釋，卷二十九，謝占王注）

This seventeenth-century text evidences the perception of distinctive local shipbuilding traditions existing along the Chinese coasts. Shipbuilders in the northern part of China preferred to specialize in the construction of a flat-bottomed ship named for its capability for navigating sandbars. In contrast, another type of ship was constructed in the Fujian and Guangdong regions in the south, using a keel (or dragon-spine) that made ships maneuverable in rocky and reef-strewn waters.

The *Wu Bei Zhi* is among the Chinese texts most cited by naval historians. This text mentions the people called Sha-min (沙民), who settled along the coast of China and excelled in naval battles with their ships, the *sha-chuan* (literally, "sand ship," 沙船), which are compared with the *Fu-chuan* ("Fu ship," meaning the Fuzhou ship, 福船), the *cang-chuan* ("blue ship," 蒼船), and the *naiowei-chuan* ("bird-tail ship," 鳥尾船). The text describes how the sha-chuan was able to tack against a headwind and states that it was convenient in the shallow waters of the northern sea but was hard to sail in the deeper southern seas. The bottom of the sand ship was flat, which did not allow it to plow through the big waves of the deep seas. There are swells in the northern sea, and the Fu ship and blue ship each had an acute angle at the bottom because of the swells, but the sand ship was also capable of sailing against these swells. In the northern sea, iron anchors could be used for mooring, but in the southern sea, the deep water allowed only the dropping of wooden anchors.

Needham (1971: 429) states that Hangzhou Bay around 30 degrees north latitude is the boundary between the two different types of ships. North of Hangzhou Bay, the coastal and seagoing craft had a flat bottom, and their suspended rudder could be lowered well below the bottom of the ship or lifted up to function in the shallow estuarine harbors and even in inland waters. The flat-bottomed ships were suitable for dual use in both the sea and inland waters. Again I want to emphasize that the impact of historical development of riverine infrastructure in the north on the flat-bottomed ships may need to be considered as important as the northern sea conditions. Intensive study of ports and possibly inland waterway navigation would probably be useful on this question, but these matters are beyond the scope of this book.

Unquestioning interpretation of geographical differences as being inseparably linked to the vessels' flat, rounded, or V-shaped bottoms may lead to an overgeneralization of the typological approach. A limited view of the variations involved is likely to suggest that the sha-chuan (sand ship) was the dominant type of flat-bottomed ship, which has been defined in comparison with the rounded and V-shaped hull. In addition, the flat-bottomed ship is likely to be characterized by no keel construction, which has been used in visual representations as a typical Chinese "junk" of the north. Despite the perception that there was diversity in flat-bottomed ships in the northern sea of the Chinese cultural sphere, most of the variation in flat-bottomed ships has been only vaguely addressed, because the sand ship has been equated with the flat-bottomed ship.

The Sand Ship Interpreted as the Jiangsu Trader

The typical image of the sand ship is most likely to have come from a drawing in the seventeenth-century historical text *We Bei Zhi* (figure 3.3). Needham (1971: 429) captions the ship depicted in this historical text as the sand ship of the Chiangsu (Jiangsu) style. George R. G. Worcester (1971) also regards the Jiangsu (Kiangsu) trader as a sand ship. According to his study, the Jiangsu trader was constructed in shipyards around Shanghai and was a seagoing ship that can be recognized by two distinctive features: five masts, and a false stern and projecting gallery three meters or more in length. The Jiangsu trader is possibly classified together "with the Antung trader as one of two fundamental types of seagoing craft of the north from which have evolved various other crafts" (Worcester 1971: 162).

The sand ship may be more similar to the Nanjing ship drawn in an iconographic resource and the Hangzhou Bay trader recorded in an ethnological boat study. Osamu Oba (1974, 1980) introduced an accurate image of a flat-bottomed ship identified as a Nanjing ship (figure 3.4) derived from the historical picture scroll *Tosen no Zu* (Images of Chinese ships, 唐船ノ図 [figure 3.5]); he labeled it a sand ship, or sha-chuan. The Nanjing

Figure 3.3. Image of a *sha-chuan* depicted in the seventeenth-century text *Wu Bei Zhi*. (From Xi et al. 2004)

ship depicted in the *Zoho Kai Tsushoko* of 1708 by Joken Nishikawa is also a useful resource to help identify the sand ship (Oba 1980). Importantly, the Nanjing ship in the iconographic resources shows similarities with flat-bottomed ships defined as the *shaohing chuan*, or the Hangzhou Bay trader (figure 3.6), in previous studies of Chinese ships in the premodern period (Donnelly 1930: 62–65; Worcester 1966: 28–32, 1971: 184–86; Sokoloff 1982: 19). Worcester gives a detailed explanation about this type of ship. The Hangzhou Bay trader had a flat bottom with no keel construction and was modified to work in the shallow waters of the bay and the shifting mudbanks of the Yangtze Estuary. It was specifically designed for riding the world's largest tidal bore.

Figure 3.4. Merchant ships depicted in the historical picture scroll *Tosen no Zu*. *Top*: Nanjing ship (*left*) and Ningbo ship (*right*). *Center*: Ningbo ship anchoring (*left*) and Fuzhou-built ship that departed from Nanjing (*right*). *Bottom*: Taiwan ship (*left*) and Guangdong ship (*right*). (Courtesy of Matsuura Historical Museum; photo by author)

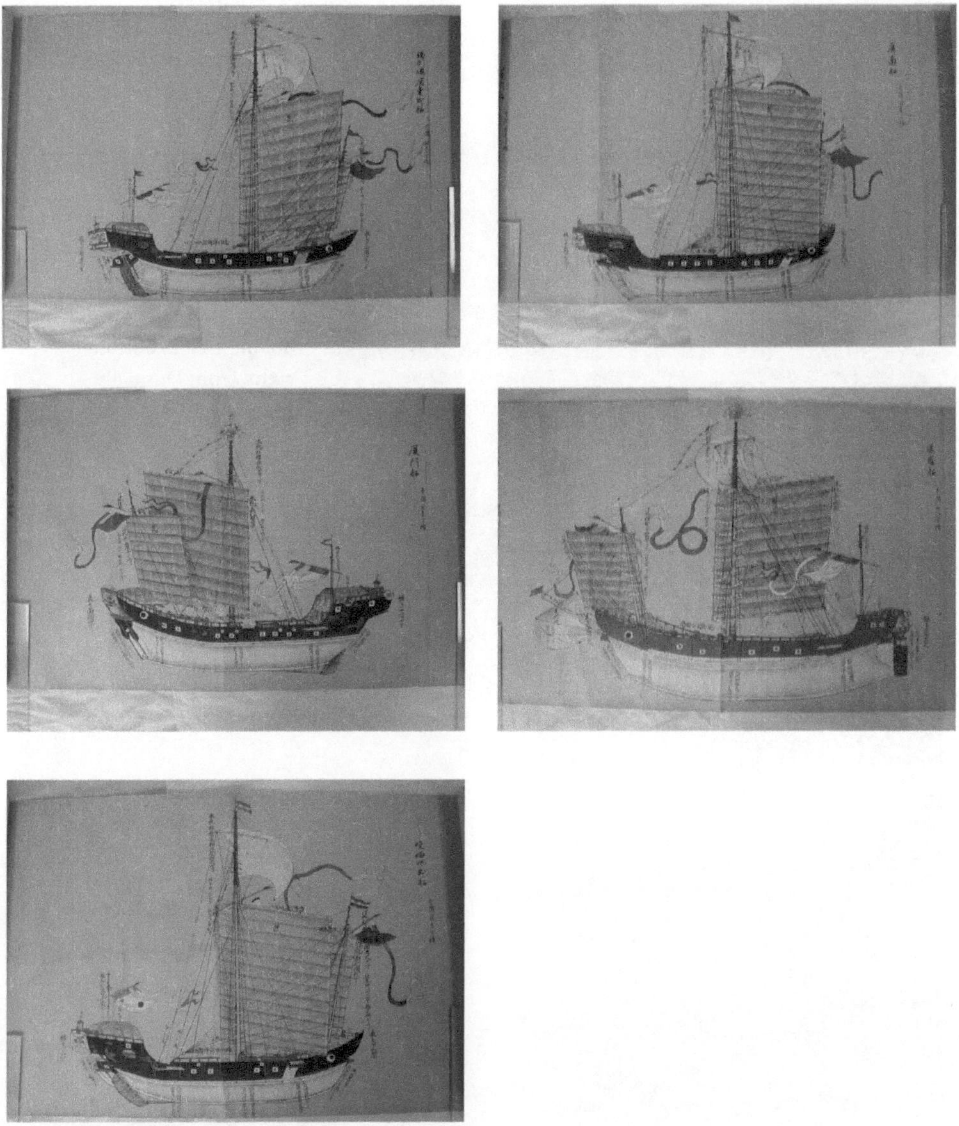

Figure 3.5. More merchant ships depicted in *Tosen no Zu*. *Top*: Fuzhou-built ship that departed from Guangdong (*left*) and Guangnan ship (*right*). *Center*: Amoi/Xiamen ship (*left*) and Siam ship (*right*). *Bottom*: Calapa/Batavia ship. (Courtesy of Matsuura Historical Museum; photo by author)

Figure 3.6. From top: Nanjing ships depicted in *Tosen no Zu* (*top*) and *Kiyo Tojin Yashiki Zukei* (*center*) and drawings of Hangzhou Bay traders (*bottom*). (Top, courtesy of Matsuura Histori-cal Museum, photo by author; center, courtesy of Kyoto University Li-brary; bottom left, from Sokoloff 1982; bottom right, from Donnelly 1930)

The Hangzhou Bay trader is characterized by a vivid painting of a stylized tiger on the bow and the adoption of leeboards consisting of two or more solid planks (Sigaut 1960; Doran 1967; Sokoloff 1982). These characteristics can also be observed on the Nanjing ship in many historical iconographies (Oba 1974, 2003). The iconographic resources depict leeboards, which are also the most definitive characteristic of ships classified as flat-bottomed ships, including the sha-chuan (sand ship), the Nanjing ship, and the Hangzhou Bay trader. Whether the Hangzhou Bay trader as recorded in ethnological boat studies had a direct link with the sha-chuan as described in the iconographic and historical resources is still disputable. These sources have limitations for assessing the detailed interior structure and construction, but we can at least see the uniformity of the shipbuilding tradition widely used around Hangzhou Bay and farther up the northern coast, including vessels used in many estuaries and inland rivers.

The Flat-Bottomed Ship in Zheng He's Fleet

The shipbuilding industry in the north around Hangzhou Bay had been undergoing development until the early fifteenth century. This is supported by the fact that some of the ships used for the historically well-known Chinese maritime expeditions during the Ming dynasty (AD 1368–1644) were constructed in shipyards at a latitude of 31°58'01" N, northwest of Nanjing and close to the Jia River, a small tributary of the Yangtze. This shipyard, called the Baochuanchang (Treasure Shipyard), could have been a significant place where some ships constituting the expedition fleet were built, possibly including the flagship of the Ming admiral, Zheng He. Relevant literature provides reasonable arguments on the issue of the size of Zheng He's ship (Sleeswyk 1996; Gould 2000: 197–98; Church 2005, 2008). The numerical information in Chinese historical texts must be carefully studied, and with regard to the dimensions of Zheng He's ship, bias and exaggerated estimates are considerable.

There is another disputable issue: whether the ships built at the Treasure Shipyard were of the northern type of sha-chuan (sand ship) or the southern type known as the Fuzhou or Guangdong ship, which had a round bottom. Xi and colleagues (2004: 208) concluded that the presence of the fleet's flagship in expeditions on the high seas in Southeast Asian waters supports the idea that large numbers of the southern type of ships were employed. However, the shipyard is geographically located in the sphere of the northern shipbuilding tradition. The long-term establishment of the Nanjing shipyards in the north could mean that they were more specialized in constructing flat-bottomed ships, and this should not be overlooked in discussion of the renowned fleet.

Comparable dimensional information about the early seventeenth-century seagoing ships from different regions of China—including Nanjing, Ningbo, Fuzhou, Taiwan, Guangdong, Guangnan, and Xiamen—is available in the *Tosen no Zu* (see figures 3.4 and 3.5). Table 3.1 shows the dimensions of these ships, as recorded in the picture scroll with a Japanese unit and calibrated into the metric system. The flat-bottomed ship built in the Nanjing region could have been relatively large (Oba 1974: 357). The recording of these dimensions demonstrates an interest in the size of the Nanjing ship, which is approximately thirty-four meters long (a few meters longer than other Chinese ships of southern origin). The dimensional data of the ships in the *Tosen no Zu* do not provide an exclusive standard by which to address the size of ships used as seagoing traders and early Ming ships but do provide a reliable illustration of ships from Nanjing's shipyards. Despotic government policy by the Ming dynasty probably affected shipbuilding industries, and some shipyards operated under government control and standardized regulation. Flat-bottomed ships constructed for the imperial fleet could be unique and be sufficiently large to attain the dimensions of the ship used as a flagship, but that ship was not built substantially beyond the uniform shipbuilding tradition. The size of the ship constructed is consistent with the size of available space at the Treasure Shipyard (Church 2005, 2008). This matter can also be logically addressed by looking at one large rudder stock found at the shipyard. The rudder stock appears to have been used as an axial rudder that could be hung on the stern; its remaining portion is 11.07 meters long. The *Tosen no Zu* provides the proportional ratios of the hull and rudder size of a Nanjing ship; the Nanjing ship shown on the scroll as having a length of 34 meters could carry a rudder consisting of a stock about 6.4 meters long. This information indicates that the rudder stock from the Nanjing shipyard could have been used for a larger ship.

For the Ming dynasty in China, which represented the rebirth of the Han Chinese after driving out the Yuan dynasty descended from the Mongol Empire, there would have been no inconvenience in using the traditional type of flat-bottomed ships for the great expedition far toward

Table 3.1. Dimensions of Chinese ships recorded in the *Tosen no Zu* (in meters)

Part of ship	Dimension	Nanjing ship	Ningbo ship	Ningbo ship (anchoring)	Fuzhou ship[a]	Taiwan ship	Guang-dong ship	Fuzhou ship[b]	Guangnan ship	Amoi (Xiamen) ship	Siam ship	Calapa (Batavia) ship
Hull	Length	34.09	29.48	31.36	29.30	29.72	29.70	29.70	30.00	32.06	42.36	29.45
	Height of bow	2.73	7.18	7.88	5.30	8.24	6.90	6.76	6.60	7.39	9.09	5.45
	Height of stern	6.82	7.18	7.88	5.30	6.90	6.90	6.76	6.60	7.39	9.09	5.45
Lower part of hull	Forward width	3.36	4.63	4.33	4.06	5.03	5.03	?	4.70	4.15	6.06	4.24
	Forward depth	1.93	2.36	4.18	2.79	2.88	3.24	3.03	3.12	3.00	5.15	2.88
	Width of midship	3.72	6.12	6.58	5.85	7.27	6.06	5.85	6.73	7.30	8.64	5.90
	Depth of midship	1.93	2.42	4.24	2.88	2.88	3.33	3.03	3.12	3.03	5.21	2.88

(*continued*)

Table 3.1—*Continued*

Part of ship	Dimension	Nanjing ship	Ningbo ship	Ningbo ship (anchoring)	Fuzhou ship[a]	Taiwan ship	Guang-dong ship	Fuzhou ship[b]	Guangnan ship	Amoi (Xiamen) ship	Siam ship	Calapa (Batavia) ship
	Aft width	5.45	5.24	6.09	4.64	6.12	5.52	4.90	5.81	6.42	7.27	6.06
	Aft depth	2.42	2.45	5.51	4.00	3.94	4.85	4.09	4.18	4.09	6.87	3.97
Transom bow	Height	3.57	3.72	3.72	3.73	4.06	4.09	3.76	4.24	4.33	5.45	4.06
	Width	2.42	2.48	2.48	1.97	2.21	2.21	2.06	2.27	2.30	3.30	4.03
Transom stern	Width	4.39	4.39	5.45	4.48	4.09	4.85	4.24	4.51	5.15	5.97	3.94
	Height	9.27	6.90	8.24	3.73	6.09	6.21	6.06	6.15	7.88	7.03	4.85
Mainmast	Length	22.54	24.70	26.21	23.33	26.36	29.24	31.06	31.97	25.70	36.45	30.42
	Circumfer-ence	1.76	2.60	2.27	2.58	2.15	2.70	2.12	2.03	2.18	2.88	2.63
	Circumfer-ence of tip	0.78	1.06	0.87	1.12	0.87	1.09	0.88	0.81	84.85	1.06	0.97

Foremast	Length	16.64	17.52	19.51	15.76	18.24	18.20	17.12	18.58	?	22.72	19.88
	Circumference	1.36	1.27	1.33	1.21	1.30	1.39	1.12	1.15	1.21	1.81	1.39
	Circumference of tip	0.54	0.54	0.51	0.51	0.54	0.60	0.48	0.48	0.45	0.75	0.54
Ensign	Length	10.90	9.55	8.79	7.67	7.72	7.58	7.73	7.73	9.85	18.45	7.73
Mainsail	Length	16.52	14.85	15.15	14.60	15.00	13.64	14.55	16.45	14.76	18.18	14.39
	Width	12.21	13.18	14.39	12.85	12.72	12.58	12.97	13.33	12.58	16.51	12.27
Foresail	Length	10.58	8.42	7.73	9.09	8.76	8.03	8.18	8.24	10.76	11.06	10.97
	Width	6.36	6.67	5.90	5.60	5.85	6.06	5.90	6.39	6.52	6.21	7.27
Keel	Length	24.64	23.12	N/A	22.21	20.06	21.15	21.45	21.72	21.55	37.72	N/A
Topsail	Length	N/A	7.06	N/A	N/A	N/A	7.58	9.15	7.88	N/A	9.33	N/A
	Width	N/A	4.21	N/A	N/A	N/A	5.60	6.00	5.85	N/A	5.90	N/A
Spritsail	Length	N/A	N/A	N/A	N/A	N/A	N/A	N/A	N/A	N/A	16.06	N/A

[a] Built in Fuzhou, departed from Nanjing.
[b] Built in Fuzhou, departed from Guangdong.

the Western world. The ships show a substantial presence in Chinese maritime history, especially in the tradition that firmly existed in the north, as illustrated in numerous historical and iconographic resources (and reviewed in previous ethnological boat studies).

The development of shipbuilding technologies can be assessed in terms of how structure and construction methods evolved historically as a tradition and made the flat-bottomed ship dominant as a representative type of seagoing vessel along Chinese coasts. The integrated approach with a focus on archaeological data in this chapter at least leads to identification of the ancestral form of the northern tradition, which is important for understanding the chronological and spatial dominance of one of the Chinese shipbuilding traditions.

4

The Quanzhou Ship

This chapter provides detailed descriptions of one of the two representative East China Sea traders discussed in this book. The remains of the hull known as Quanzhou Guchuan (Quanzhou Ancient Ship, 泉州古船), or the Quanzhou ship, have been studied over three decades as one of the most important and intact archaeological shipwrecks ever discovered in East Asia. The Quanzhou ship is a Chinese trader dating to the second half of the thirteenth century. A comprehensive report, the collaboration of many Chinese researchers, was published by the Museum of Overseas Communication History (1987), which is the custodian of the ship remains. The Quanzhou ship has been reviewed and assessed in several English-language studies (Merwin 1977; Keith and Buys 1981; Green 1983a; Steffy 1994; Burningham and Green 1997; Green et al. 1998; McGrail 2001). The ship is currently displayed at the Quanzhou Ancient Ship Gallery in Kaiyuan-shi (Kaiyuan Temple, 開元寺) at Quanzhou city in Fujian Province. Because of preservation problems, the condition of the ship remains is somewhat different from that at the time of its discovery in the 1970s. In this chapter I review the in situ status of the hull while it was being excavated, based on original reports and photographs produced from the excavation work. Further details about the structure of the hull have emerged from research conducted by members of the Department of Maritime Archaeology at the Western Australian Museum. Also included are the results of my archaeological inspection of the Quanzhou ship in 2008 and 2009. By reviewing previous work and giving a new perspective, I aim to clarify the structural characteristics of the Quanzhou ship's hull and the principles of the construction method that were the conceptual basis of the shipwrights' work during the construction of this ship.

Site Background and the Ship's Discovery

The Quanzhou ship was the first vessel archaeologically excavated in China or the East Asian region. Its discovery raised marked public interest in local history and evoked the heyday of Chinese maritime trading. Since the discovery, the Chinese archaeological journal *Wenwu* (文物) has published a series of articles on the excavation and research. English translations of these articles were later produced, resulting in the first resource (Merwin 1977) available for scholars not reading Chinese.

The Quanzhou ship was found in the tidal mud flats of the modern Houzhu Harbor along the river Luoyangjiang flowing into Quanzhou Bay (figure 4.1). Encompassing large estuaries of two major rivers, the Luoyangjiang and Jinjian, Quanzhou Bay has been known as a mooring place throughout history. The bay contains several ports, including Houzhu and Fashi. These harbors rose in prominence from the tenth to the thirteenth century as major ports serving the city of Quanzhou (called Zaiton at the time). The area of the ports, approximately ten kilometers away from the city of Quanzhou, was known as Linjiangli during the Song dynasty (AD 960–1279), and in the thirteenth century during the Yuan dynasty (AD 1271–1368) it was renamed Houzhu. Throughout the Song and Yuan dynasty periods, the area was settled by foreign merchants. The growth of international trade is evidenced by various multicultural and religious monuments remaining in the Quanzhou region (Chaffee 2001; Pearson et al. 2002). Despite flourishing as an international port, Quanzhou Bay became less significant during later periods because of environmental factors affecting the harbors, including the accumulation of silt. After the fourteenth century, other historical cities such as Amoy began to dominate the international seaborne trade, although Houzhu Harbor still functions as a major port on Quanzhou Bay in modern times.

The Quanzhou ship was found in 1973 during dredging in the southwestern part of Houzhu Harbor, below sea level and covered by sediments 2.1–2.3 meters thick (figure 4.2). According to Douglas Merwin (1977: 10), "When the vessel was discovered, only a portion of its side was exposed." The site plan indicates that this was the starboard side of the hull (Museum of Overseas Communication History 1987: 7). A test excavation was implemented in the year of the discovery. As a result of the test excavation near where the Quanzhou ship was found, stone pilings and wooden platforms were also discovered, presumably associated with an ancient slip or

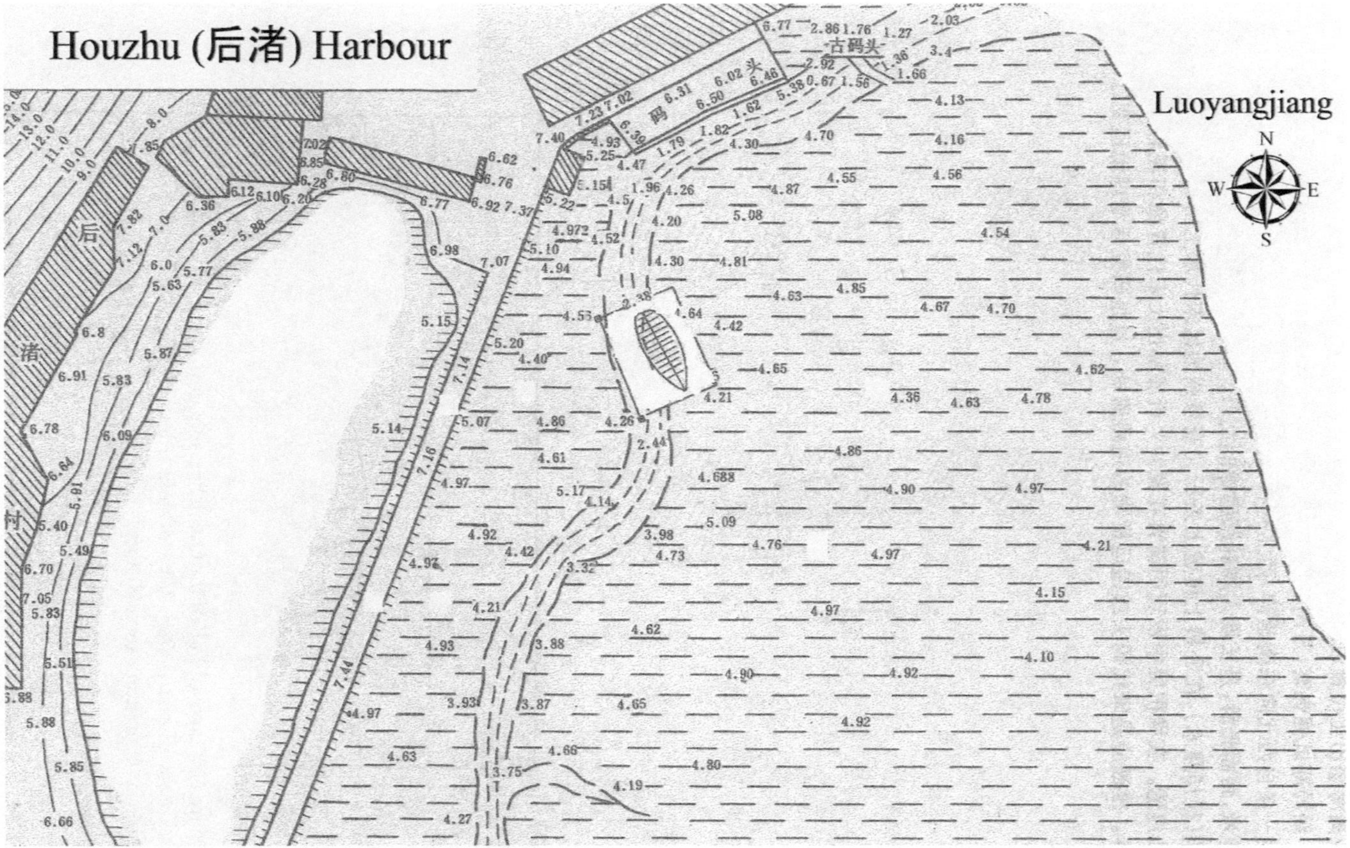

Figure 4.1. The Quanzhou ship's location in Houzhu Harbor. (After Museum of Overseas Communication History 1987)

Figure 4.2. Exposed hull of the Quanzhou ship during excavation. *Top*: View of hull looking southeast. *Bottom*: View of stern from port side. (Courtesy of Museum of Overseas Communication History)

part of the harbor structure, though details of the relationship between this maritime infrastructure and the Quanzhou ship were not clearly explained (Merwin 1977: 10). The full excavation started in 1974 and continued for a month and a half. The local authorities involved included the Chin-chiang District Revolutionary Committee, the district's Association of Cultural Organizations, the History Department of Amoy University, and the Museum of Overseas Communication History.

The excavation initially removed thick sediments covering the hull. The strata were recorded and classified into three layers. The first layer (400–600 millimeters thick) is coarse grayish yellow silt with a little sand and containing modern artifacts, such as roof tiles and ceramic sherds. The second layer (600–700 millimeters) is light gray-blue fine silt and contains ceramics dating from the Song to Qing dynasty periods; between the second and third layers is a thin sand layer. The third layer is dark gray fine and dense silt and contains Song and Yuan dynasty ceramics (Museum of Overseas Communication History 1987: 8–9). After the accumulated sediments were cleared, the hull was exposed. The quality of sediments observed inside the hull was the same as that of the third layer (Museum of Overseas Communication History 1987: 14).

A small trench was opened under the hull to assess the sediment on which the hull lay. Dense fine silt was observed just under the hull, and there was soft gray fine silt as well. In these sediments beneath the ship, ceramics dating to the earlier part of the Song dynasty were identified. In the dense gray fine silt, large pieces of oyster shells were found, and some oyster shells were also discovered on the outer surface of the bottom of the hull. Thus, the excavation report indicates that the study of sedimentation was a primary determinative factor for the relative date of the Quanzhou ship. Burial conditions of the ship in the sediments, however, have not been fully explained from the viewpoint of site formation processes. The hull was noted to be on a level plane when it was found (Merwin 1977: 11). According to Merwin (1977: 18), "Inspection of the present condition of the vessel's hull reveals virtually no signs of it having been repaired. It was a seagoing vessel that had been built and in service not long before." These descriptions are not sufficient to determine whether the ship was accidentally wrecked or deliberately abandoned.

Hull Dimensions, Form, and Basic Structure

Overall dimensional information about the Quanzhou ship is outlined here; fuller detail appears later in descriptions of the components. During the full excavation in 1974, the well-preserved lower hull was revealed, and it was dismantled to be carried away from the site. The dismantled components were reconstructed at Kaiyuan-shi (figure 4.3). The overall length of the hull remains is 24.2 meters, and the hull is 9.15 meters in breadth (Merwin 1977: 11; Museum of Overseas Communication History 1987: 16). The main structure of the hull consists of a keel, twelve bulkheads, and sturdy planking. The hull planking forms clinker-like steps and consists of multiple layers, double layers changing to triple layers around the turn of the bilge. The forward part of the hull narrows to a sharp shape (perhaps a narrow transom), compared to the broad stern (figure 4.4). The cross section of the hull shows considerable deadrise in the bow; however, toward the stern the sections are flatter and broader. Chinese researchers have estimated that the cargo capacity of the Quanzhou ship could have been more than 200 metric tons, and another study has demonstrated that the laden displacement of the Quanzhou vessel was approximately 370 metric tons (Merwin 1977: 81, 94). The suggestion has been made that the

Figure 4.3. Reconstruction of the dismantled hull of the Quanzhou ship at the Kaiyuan Temple. (Courtesy of Museum of Overseas Communication History)

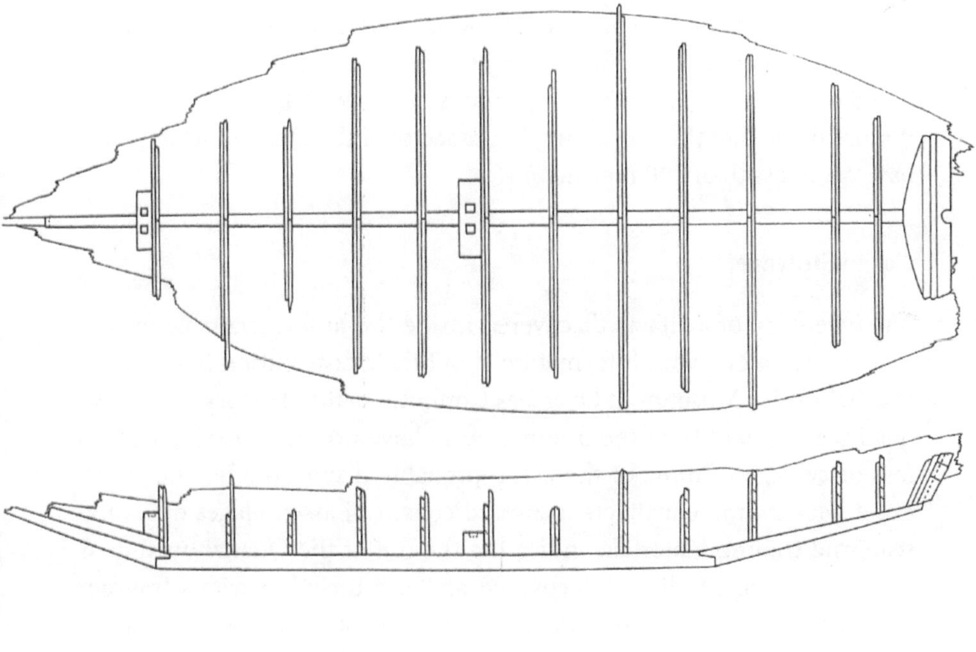

Figure 4.4. Quanzhou ship, plan of the hull and ship lines. (Courtesy of Museum of Overseas Communication History)

Quanzhou ship could have carried fifty people and some livestock (Merwin 1977: 96). However, because the upper part of the hull is missing, a precise reconstruction cannot be made. Using a comparative perspective based on study of the Shinan shipwreck's hull (see chapter 5), I consider it more likely that the cargo carrying capacity of the Quanzhou ship may have been less than 200 metric tons.

Cargo Stowage

The inventory of artifacts discovered inside the hull is available in several reports, along with information about their distribution in the holds (Merwin 1977; Museum of Overseas Communication History 1987). The quality and quantity of the original cargo have not been ascertained, as loss or salvage of most of the cargo probably occurred after the wreck event. However, the artifacts recovered consist of assemblages of several seaborne trading materials, indicating the use of the Quanzhou ship as a trader (see appendix 1). Recovered artifacts include various fragrant woods and spices, such as pepper. These were common commodities in Asian maritime trade (Schottenhammer 2010). Chinese ceramics are also typical seaborne commodities likely to be identified in Asian merchant shipwrecks. However, the amount of Chinese ceramics recovered from this hull is relatively small. Most of these artifacts can be regarded as items that were originally carried as cargo. Several inked wooden strips were found; these seem to be labels or tags associated with the trade items on the ship.

Distribution of the cargo is used to interpret the utility of the holds and the cargo loading patterns of thirteenth-century East Asian seagoing ships. In the case of the Quanzhou ship, however, very limited information about cargo distribution is available, because detailed site plans were apparently not produced during the excavation. Apart from the trade goods, artifacts from the hull include possible personal belongings such as carpentry tools, chopsticks, and ivory chess pieces. These appear to be from the aft holds—the twelfth and thirteenth holds. The wider distribution of pieces of fragrant woods, from almost all holds apart from the thirteenth (the aftermost) hold, indicates that they might have been a part of the main cargo. These fragrant woods, however, show relatively small dimensions, measuring 30–100 millimeters on average (the largest piece is 1.68 meters in length). In the case of the Shinan shipwreck (see chapter

5), the type of fragrant wood is of only one species (red sandalwood), and the pieces are longer, on average. In the Quanzhou ship, each hold space is smaller than in the Shinan ship. Therefore, the design of the Quanzhou ship's stowage space was more suitable for smaller cargo items.

Details of the Hull Structure

An intensive suite of surveys highlighting structural and construction aspects of the Quanzhou ship was conducted by a research team from Australia between 1983 and 1994 (Burningham and Green 1997; Green et al. 1998). The results of the four surveys, performed in cooperation with local researchers, included a detailed plan of the ship lines and a plan of the hull structure drawn from offset measurements, careful visual observation, and the use of an electronic distance-measuring system in 1994 (figure 4.4). In 2008 and 2009, I further inspected the site for more a detailed hull analysis. These most recent assessments were aimed at recording further details of the hull structure.

Since the excavation in 1974, the hull of the Quanzhou ship has deteriorated so that physical contact must be minimized, and access to the hull was limited during my latest survey. The deterioration appears to be the result of the relative lack of conservation methods that were available when the hull was recovered from the estuary. According to Guoqing Li (2004), the intent was to season the waterlogged timbers as much as possible by slow drying, and conservators spent four years completing the drying process with treatments that included covering the ship remains with plastic sheets to retain moisture, disinfection of the timbers by spraying a mixture of water and 5 percent alcohol, and control of humidity in the storage room. To prevent exfoliation of the timber surface, coatings consisting of sesame oil, tung oil, and acetone were applied. For preservation of the timber, a substance made from resin and paraffin wax was used. Despite these efforts, deterioration appears to have continued. Structural distortion of the hull, including shrinkage and cracking of the planks, is visible. An example of changes in the hull can be seen in the stern transom, comprising a vertical set of three layers of solid timbers; in stark contrast to the current condition of the stern, this was originally well preserved, and a circular rudder socket was visible in photographs (figure 4.5). Information regarding the original dimensions of the rudder socket appears to have been lost. The photographic record suggests that

Figure 4.5. Stern of the Quanzhou ship at present (*top left*) and as recorded in a black and white photograph taken during the excavation (*top right*). Modern iron nail seen at the bottom of the hull (*bottom*). (New photos [top left and bottom] by author; older photo [top right], courtesy of Museum of Overseas Communication History)

there could have been more accurate data recording immediately after the hull was revealed. Even from the remains of the stern in their current condition, however, it is still obvious that the rudder was raked at the same angle as the transom and that the rudder post could slide up and down in the socket.

Further, many iron nails were initially used for the modern hull reassembly, and later their corrosion became the cause of secondary damage, resulting in additional recent deterioration of the timbers. In an attempt to find a fastening system that would have less impact on the hull, conservators replaced the iron nails with bamboo nails (Li 2004). But this work seems to have been incomplete, and many corroded iron nails as well as bamboo nails are exposed in the lower part of the hull (figure 4.5). In addition to the degradation of the nails themselves, many cracks are evident around these iron nails and the holes where they penetrate the hull planks.

A more serious consideration is the position of the modern nails, which are said to have been driven into the original nail holes. In the outer structure of the hull, Portland cement was used to seal the heads of the modern nails or perhaps to fill both original and nonoriginal nail holes. Nails were driven from the surface of the outermost planking to the inner planks, yet the nails' driven patterns appear more or less random. Whether these modern nails were located exactly in the original nail positions could not be examined, because of the lack of recorded details of the reconstruction work. The structural distortion of the hull and imprecise reconstruction have been taken into consideration here to allow an acceptable range of error in recovering data from the ship. Furthermore, in an attempt to minimize the risk of misinterpretations arising from reliance on information from the current ship remains, the assessment in the following section of the structural characteristics of the Quanzhou ship is made partly from the original resources—reports and photographs produced in the 1970s and 1980s—as well as the results of the more recent set of surveys.

Keel

The keel of the Quanzhou ship is composed of three parts: a forward keel, a main keel, and an aft keel. These are joined by stopped scarfs with tenons (figure 4.6). In the Chinese shipbuilding tradition, a keel is called a *longwu*, meaning a dragon spine (see glossary). Accurate dimensions and the configuration of the keel in the current hull are elusive because of the difficulty of access to the inner hull. However, dimensions and details are provided in the existing literature. The overall length of the keel (consisting of the main keel and aft keel) is 17.65 meters, and its width is 0.45 meters (Merwin 1977: 11–14; Museum of Overseas Communication History 1987). The depth of the keel is about 400 millimeters (the depth of the keel

Figure 4.6. Scarf of the keel showing circular holes in which coins and a mirror could be placed. (Courtesy of Museum of Overseas Communication History)

below the garboard, which sits down in a rabbet, is 270 millimeters). The length of the main keel measures 12.40 meters, and the length of the aft keel is 5.25 meters. The aft end of the aft keel is tapered. While the forward keel measures 4.50 meters, the end of the forward keel may be partly lost. The forward end of the forward keel is tapered as well and measures 180 millimeters in width and 200 millimeters in height.

The forward and aft keels are scarfed to the main keel and are angled upward; the forward keel slopes upward 20 degrees, and the main keel 15 degrees. In the joints or scarfs, the upper members of the forward and aft keels overlap the lower member of the center keel by fitting an end joint with an added mortise and tenon. Remarkably, there is no recorded evidence of nails being used on these scarfs (Museum of Overseas Communication History 1987). A tradition of placing seven coins or coin holes with a bronze mirror, known as *bao shou kong* (see glossary), was confirmed in the scarf between the forward keel and main keel.

Structure of the Bow and Stern

Upper structures of the bow and stern of the Quanzhou ship are missing, but their characteristics can be partly inferred from the remains of the lower hull and the general characteristics of Chinese watercraft. Modern reconstruction at the bow prevents direct examination of the original configuration (figure 4.7). A large timber was apparently placed forward of the keel scarf of the forward and main keel, according to the plan of the ship (see figure 4.4). It is regarded as a hog piece. A forward mast step sits

Figure 4.7. The Quanzhou ship's bow (*top*, front view) and stern (*bottom*, aft view). (Photos by author)

on the aft end of the hog piece. The first bulkhead is placed immediately abaft the aft end of the hog piece and mast step. The fastening methods of these components could not be examined, as their joins are not visible in the reconstructed hull. With regard to the missing part of the forward keel and bow structure, whether the Quanzhou ship had a trapezoid-shaped bow transom is open to question. In the case of the Shinan shipwreck, a transom bow surmounted the end of the forward keel (see chapter 5). A similar structure may have been used on the Quanzhou ship.

Three layers of heavy timber baulks appear to compose a stern transom, barely identifiable given the current state of the hull. The layers are not evident on an old photograph of the stern (see figure 4.5). A layer of thin sheathing can be seen on the outside of the current stern transom, and this is evident in the old photograph (see figure 4.5). The published information indicates that the remaining part of the stern transom measures 1.37 meters (Museum of Overseas Communication History 1987: 21). The breadth across the top of the remaining transom is 3.44 meters, and the width of the two or three baulks is 0.44 meters. The stern transom is raked aft 20 degrees. A large slot for a rudder stock passes through the center of the aft face of the transom. The diameter of the slot is 380 millimeters. This is slightly less than the width of the baulk. The current dimensions of the slot may not precisely reflect its original dimensions, given the deterioration of the timbers around the hole. There is no clear evidence as to how the rudder stock was secured to the stern transom. The rudder post could simply have been slid into the slot. The rudder was probably similar to the hoisting type of rudder as recorded in Worcester's ethnological boat study of Chinese junks (Worcester 1971: 98). A structure to retain the rudder of the Quanzhou ship is designed for a transom-mounted rudder. In the same way as premodern junks used wooden gudgeons (or sockets), the rudder of the Quanzhou ship could have been slung from a windlass and raised or lowered as circumstances required. Hull planking in the stern extends beyond the stern transom to form a sort of false counter stern. This indicates that a superstructure including a stern gallery could partly have shielded the stern transom as well as the rudder.

Hull Planking

Hull planking of the Quanzhou ship consists of double- and in parts triple-layered planking. The strakes are edge-joined by skew nailing through

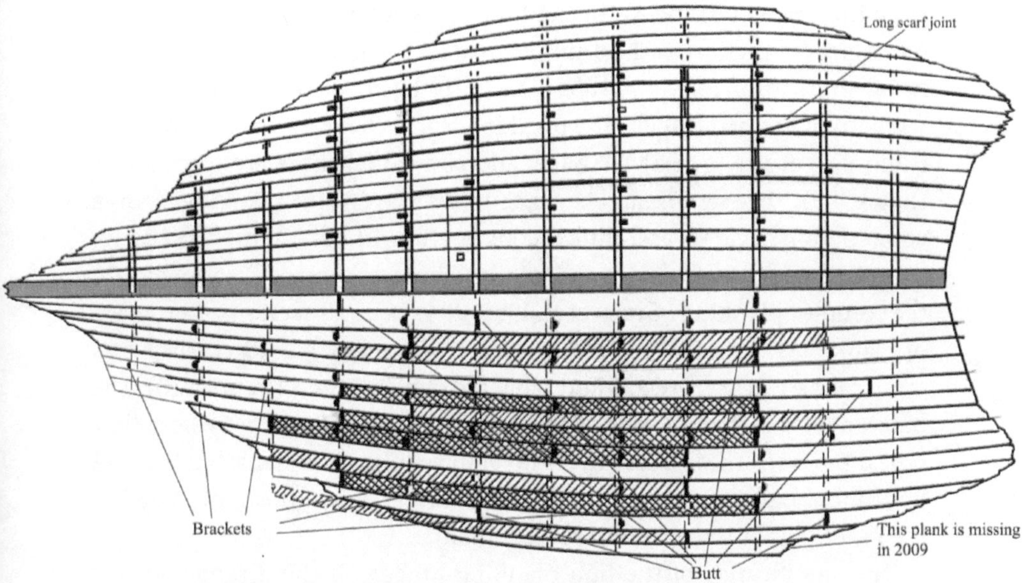

Figure 4.8. Plan of the planking arrangement, showing the positions of the iron brackets and butt. (After Burningham and Green 1997)

rabbets, which produce clinker-like steps in some strakes, while most are flush or carvel. This skew-nailed and rabbeted construction is important to the strength of the hull. Thirteen strakes remain on the starboard side and fifteen strakes remain on the port side. During the survey in 1994 a sixteenth strake was recorded on the port side (figure 4.8). However, in the 2009 survey the uppermost strake (the sixteenth) was observed to be missing, and the uppermost sheathing plank associated with the fourteenth and fifteenth strakes had also disappeared. Each strake consists of two or three planks per side, their butts showing two different joint methods: a short lap joint and long scarf joint (figure 4.8). However, the long scarf is an exception, appearing in only one join. The lengths of the planks range from 9.21 to 13.50 meters. Their widths are 280 to 380 millimeters. The thickness of the planks shows a difference between the layers. The first, inner layer is about 82–85 millimeters in thickness, compared with the two outer-layer planks, consisting of the second plank at 50 millimeters thick and the third plank (the outermost planking layer) at 30–50 millimeters thick.

The hull planking of the Quanzhou ship thus shows a complex structure. An earlier study has highlighted the complexity of the planking system (Green et al. 1998). In the hull planking of this ship, one to two outer layers function as sheathing to protect the main planking from the naval shipworm, *Teredo navalis*. The outer layers also contribute to the strength of the plank shell. The planking is in two layers up to around the beginning of the turn of the bilge, and above that there are three layers (figure 4.9). To assemble and key the multiple layers, the planking forms a combination of carvel and clinker construction. The terms *rabbeted carvel seam* and *rabbeted clinker seam* (see glossary) were introduced to explain the complex planking (Green et al. 1998: 284). A cross section of the layers of planking shows that the assemblage of the inner planks consists of a set of rabbeted carvel seams and rabbeted clinker seams (figure 4.10). The outer layers are fitted to the clinker steps but have no rabbet where they repeat the step. These outer layers form the clinker-like steps, but they are not functionally of clinker construction; rather, they are simply fitted to the inner planks to cover and protect them.

The significance of the hull planking hinges on the arrangement of the inner layer as follows. The rabbeted clinker construction appears on the seams of the second and third strakes, fifth and sixth strakes, eighth and ninth strakes, and eleventh and twelfth strakes, which amount to the remains of four clinker steps (figure 4.10). The rabbeted carvel seams appear between those clinker steps. The pattern of the set of rabbeted carvel seams and rabbeted clinker seams continues up to the twelfth strake. All grooves on the rabbeted clinker and rabbeted carvel seams show consistency, being cut into the inside on the upper strakes.

Iron nails are used for the fastening of the planking. Nick Burningham and Jeremy Green (1997) confirmed that these nails were diagonally driven down from the upper plank to the lower, so the fastening can be termed *skew nailing*. The nails are called *ding ting* in Chinese reports (Museum of Overseas Communication History 1987; see glossary). A Chinese scholar estimated that the nails used for the planking were 200 millimeters in length (Burningham and Green 1997: 42). With regard to the original nails, their corroded remains and driven patterns are scarcely identifiable in the currently visible part of the hull planking. Nevertheless, during the survey in 2008, efforts were made to identify original nails or nail holes. This was based on an assumption that the upper strakes, which are away from the keel, were less impacted by the modern restoration work described earlier in the chapter, such that replacement with modern nails might not have been carried out in some parts. Strake no. 13 of the inner

Figure 4.9. Three-layered hull planks showing rabbeted seams in the innermost planks. (Courtesy of Museum of Overseas Communication History)

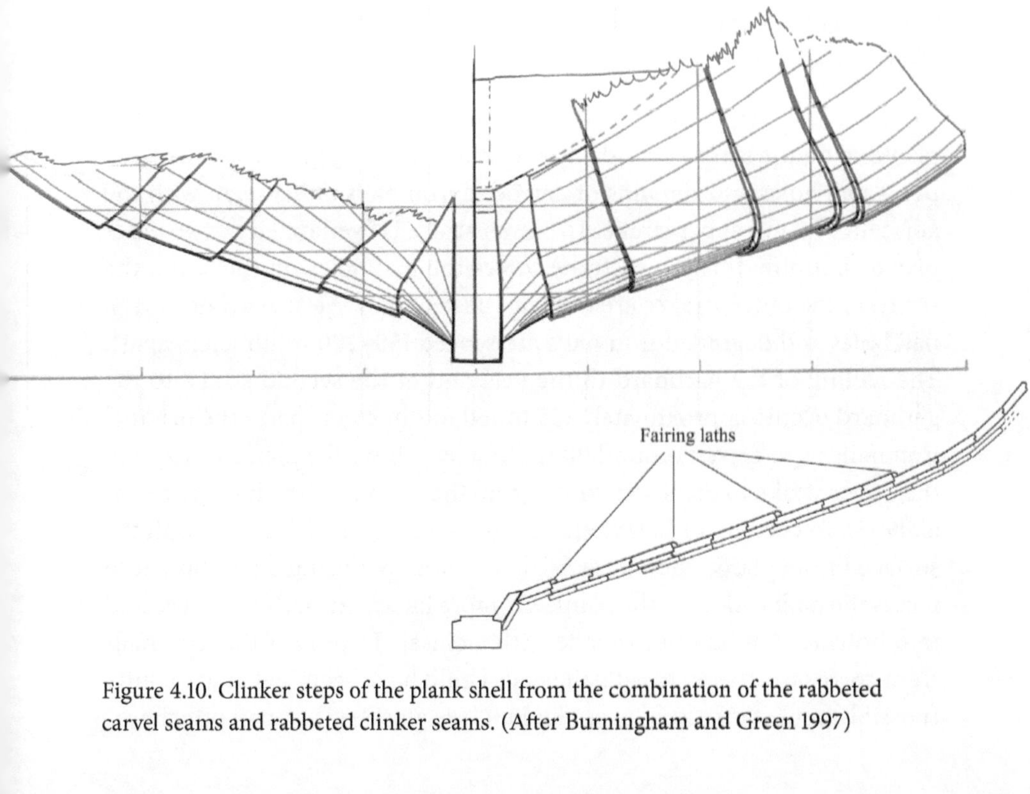

Figure 4.10. Clinker steps of the plank shell from the combination of the rabbeted carvel seams and rabbeted clinker seams. (After Burningham and Green 1997)

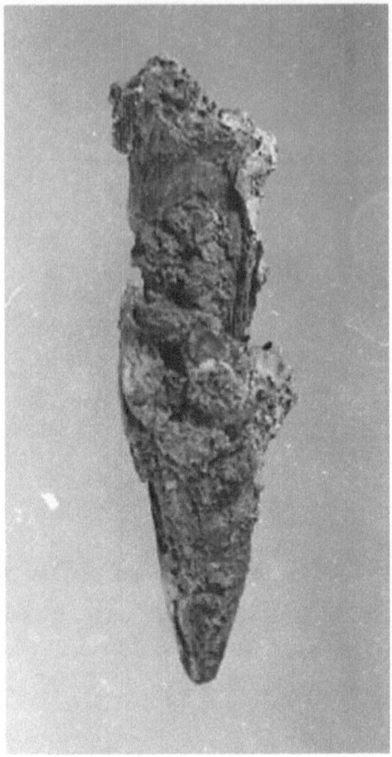

Figure 4.11. Degraded metal object found in the hull, believed to be a nail punch. (Courtesy of Museum of Overseas Communication History)

planking on the starboard side, the remnant of the uppermost strake, exposes nail holes and remains of corroded iron nails on its seam with the adjacent missing upper strake. The inner strake is overlapped by two outer planks, but only this inner strake shows the nail holes on the seam; the seams of the outer strakes are not edge-fastened by the skewed nails. The nail holes and degraded iron nails are spaced 150–200 millimeters apart. The nailing of the garboard to the keel and of the second strake to the garboard occurs approximately 125 to 150 millimeters apart. The original iron nails were driven around 90 millimeters above the plank seam from the upper strake to the lower strake from the outside of the hull (Green et al. 1998). According to Li (1989), after nails were fully driven through the surface of the plank, they were further hammered using a nail punch to recess the nailheads into the plank. Or more likely, the nails were recessed into notches that had been made with a chisel. To protect the iron nails from corrosion, the recessed nail heads could have been sealed with putty (possibly a substance known as *chunam* [see glossary]). A nail punch was found in hold no. 13 at the bottom of the Quanzhou ship. The remains of this tool, however, seem to exist no longer. There is a black and white photograph of it, but details based only on the photograph are incomplete (figure 4.11).

Bulkheads

The Quanzhou ship features bulkheads as a transverse component of the hull. Twelve bulkheads produce thirteen holds in the hull. Upper planks of most bulkheads are missing, and the remains of the planks vary in each bulkhead. For example, bulkhead no. 8 has six planks, the largest number remaining, and measures 1.86 meters in height. In contrast, on bulkhead no. 6 only three planks remain, and this bulkhead measures 0.80 meters in height (figure 4.12). These bulkheads were reported to be irregularly spaced (Merwin 1977). The largest spacing, that of the eleventh hold (between bulkheads nos. 10 and 11), measures 1.80 meters; for the twelfth hold, where the narrowest spacing occurs, the space between the bulkheads measures 0.80 meters. It is also obvious that the first two and last two bulkheads are narrowly spaced. In the 2009 survey, I reexamined the bulkhead spacing. The result shows more regularity in spacing from bulkhead no. 2 to bulkhead no. 10 than was previously recognized. These bulkheads are spaced approximately 1.5 meters apart, with variation from 1.42 to 1.55 meters. The survey revealed that some bulkheads were not correctly located in the original positions in reconstruction processes. The measured spacing was corrected for observed misalignments. For example, the upper part of bulkhead no. 1 is not as far forward as it should have been and does not rake forward enough. This prevents it from properly meeting the upper planking.

The thickness of the bulkhead planks is approximately 80–120 millimeters. Their width ranges from 250 to 530 millimeters. The topmost extant plank tier of bulkhead no. 4 consists of two planks, which are joined together by stopped scarfs. Similar scarfs also appear in the two top planks in bulkhead no. 8 and in the topmost extant plank tier of bulkhead no. 9. According to an old photograph, metal clamps could have been used to fasten these scarfs, but this cannot be seen in the current hull (figure 4.13). Apparently, the same fastening method by clamps could have been used to tighten the bulkhead plank tiers (figure 4.13). These clamps are not mentioned in previous reports, and the absence of a record of these

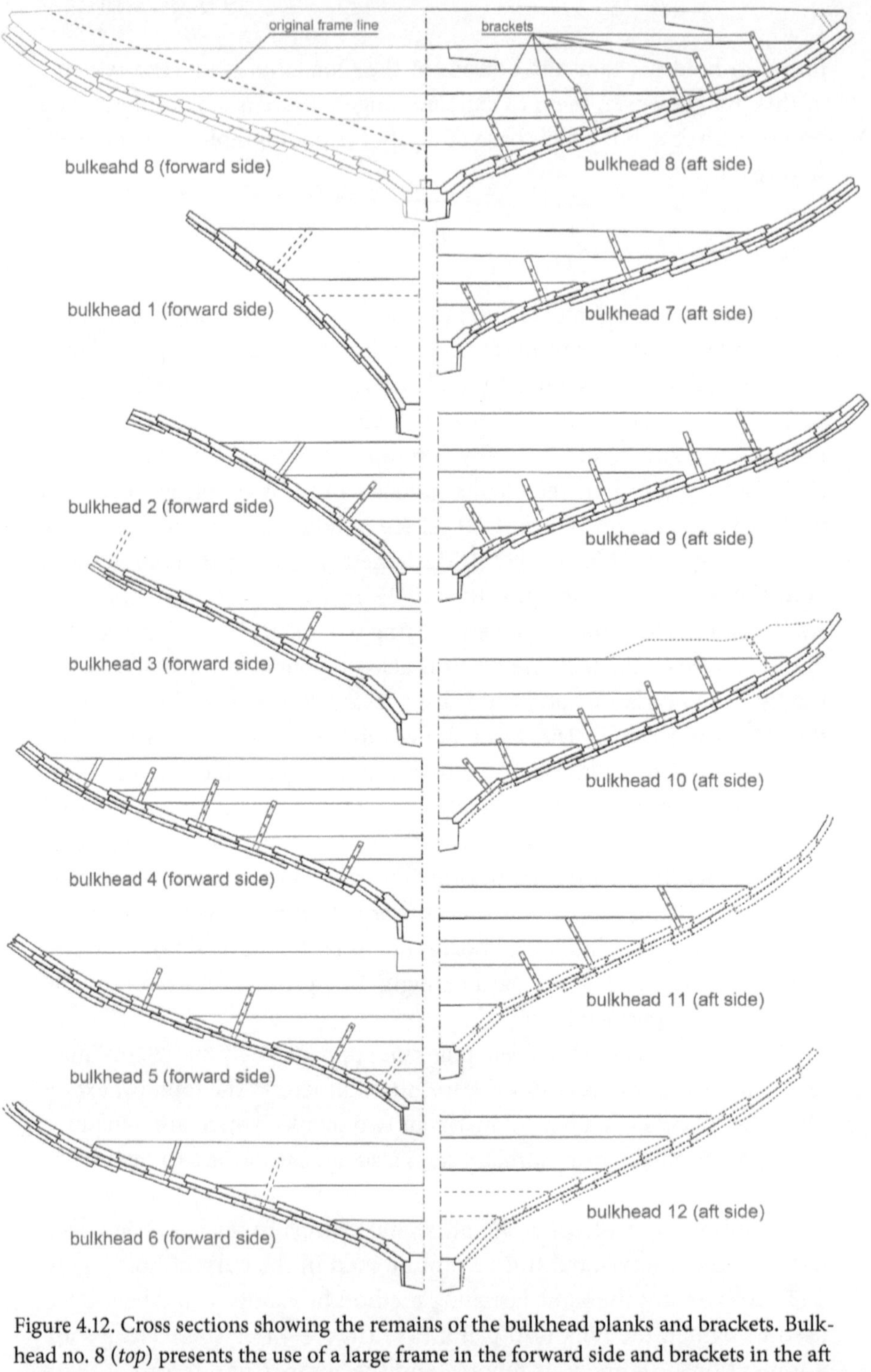

Figure 4.12. Cross sections showing the remains of the bulkhead planks and brackets. Bulkhead no. 8 (*top*) presents the use of a large frame in the forward side and brackets in the aft side, and the same arrangement appears in bulkheads nos. 7–12 (*right side images*). Bulkheads nos. 1–6 (*left*) have the brackets forward and large frames aft. (Courtesy of the Western Australian Museum; drawings by author)

Figure 4.13. Stopped scarf joint of the bulkhead plank fastened by an iron clamp (*top*) and iron clamps fastening bulkhead planks together (*bottom*). (Courtesy of Museum of Overseas Communication History)

fastenings initially aroused suspicion that they might be a modern installation. Yet the use of iron clamps or dogs to tighten plank tiers was a common technique in Chinese shipbuilding traditions (Worcester 1966). The early use of iron clamps can be carefully addressed with the possible example of the Quanzhou ship.

Upper and lower edges of the bulkhead planks are rabbeted to join the planks tightly. The groove of the rabbet is cut on the forward side up to bulkhead no. 9, and from bulkhead no. 10 on, the groove appears on the aft side. Iron nails are also used for the fastening of the bulkhead planks. These nails are driven downward from both the forward and the aft face of the bulkheads. They appear to be rather randomly spaced. The use of wooden bulkhead fastenings such as wooden tenons has not been identified in the Quanzhou ship. Because the exposed seams of the bulkhead planks are obscured, however, this detail could not be determined with certainty. The use of wooden bulkhead stiffeners, which appear to have been vertically attached to the bulkhead surface, is barely evident in the old photographs (see figure 4.12). These are missing in the current hull, though in some bulkheads the modern wooden strips have been reconstructed to look like the originals (figure 4.14).

The sides of the bulkheads have been carefully shaped. They are cut to fit neatly to both the carvel and the clinker profile of the inside of the hull. The edges of some bulkhead planks are notched or hooked to fit over clinker seams, but this is seldom the case, because the seams of the bulkhead planks often align with the clinker seams of the hull planking in a way that prevents the bulkhead planks from being hooked over the clinker step (see figure 4.12). There are fairing laths on the inside at each clinker step. The fairing laths prevent water and debris from lying in the clinker steps. Some fairing laths run under the bulkheads, while others are cut to fit between bulkheads.

Limber holes for drainage of bilgewater are cut into the bulkheads. The old photograph shown in figure 4.15 is an early record of the bottom plank of a bulkhead: the bottom plank is substantially degraded, compared with the upper bulkhead planks. The degradation appears on most of the lowest bulkhead planks from bow to stern. These bottom planks were likely kept wet by bilgewater much of the time and became waterlogged. According to a Chinese report, limber holes had dimensions of 120 × 120 millimeters (Museum of Overseas Communication History 1987: 19). Unimpeded by the bulkheads (and frames), the bilgewater could flow along

Figure 4.14. Vertical wooden strips in bulkheads nos. 10 and 11 represent the reconstruction of the original bulkhead stiffeners. (New photo [above] by author; older photo [left] courtesy of Museum of Overseas Communication History)

the channel formed by the garboards, which stand at a fairly steep angle of deadrise from the keel. Bilge boards or bottom ceiling planks sit on the seam between the bottom plank and second plank. These planks or boards would have kept cargo and other things from falling into the bilge (figure 4.15).

Figure 4.15. Bilge boards (or bottom ceiling planks) and the degraded bottom bulkhead plank. (New photo [above] by author; older photo [right] courtesy of Museum of Overseas Communication History)

Fastenings

The bulkheads are fastened to the hull planking by metal brackets called *gua ju* (see glossary). Details of the brackets in the Quanzhou ship are given in previous studies (Xu 1985; Museum of Overseas Communication History 1987: 19–20; Li 1989; Burningham and Green 1997: 44; Green et al. 1998: 288–89; Zhang et al. 2004: 288–90). The brackets were made of iron. Most of the brackets have been completely lost through corrosion and probably past salvage, but a few corroded fragments remain. The shapes and positions of the brackets are shown by iron corrosion deposits on the faces of the bulkheads. In some cases, slight recessing in the bulkheads where brackets were let in is of use in determining the original configuration. The brackets are L shaped and measure approximately 300–550 millimeters in length, approximately 50–60 millimeters in width, and about 6–7 millimeters in thickness, based on the evidence of the slits in the hull planking through which they passed. The L-shaped portion of the brackets is fastened on the outer face of the inner planking, and the brackets passing through slits in the planking are recessed into the surface of the bulkheads. The depth of this recess varies, and in some cases the recess is slight or absent. The brackets are fixed on the planks and bulkheads by four or five short iron nails. These iron nails are only barely identifiable in the current hull. At one place, the nail seems to have been driven diagonally, not perpendicularly, into the bulkhead. This is an inconvenient way to fix the brackets, and a single example cannot be relied on as indicative of the normal arrangement. As to how the brackets were fitted, hull planks and bulkhead planks were chiseled to make slits and recesses before the brackets were installed. In the uppermost plank of bulkhead no. 8 on the port side (and possibly on the starboard side as well), there is a very slight recess (figure 4.16). This might be evidence that the bracket was wrongly positioned at first and that the shipwright stopped the chiseling.

The distribution of the brackets on the hull planking shows a logical pattern and some regularity (Burningham and Green 1997: 44; Green et al. 1998: 289–91). As mentioned, two or three planks make up each strake of the hull, and the brackets are never positioned at the butts of the planking but instead are almost all adjacent to the butts (see figure 4.8). The position of the brackets is symmetrical on the port and starboard sides except for one extra bracket in the tenth strake that is positioned at bulkhead no. 8 on the port side only. Not all strakes are fastened to each bulkhead with

Figure 4.16. Part of a bulkhead plank (*inside the circle*) that shows evidence of recessing not completed. (Photo by author)

brackets. There are no brackets fastening the garboard strakes, as they would not fit, but all other substantially intact strakes have at least one.

Forty-six bracket positions on the strakes on the port side were recorded in the 2009 survey. On the fifth and eleventh strakes, only one bracket was confirmed. The seventh strake shows the greatest number of brackets: six. The number of brackets on the other strakes ranges from four to five. The number of brackets remaining on each bulkhead on the port side (some indicated only by the recesses) is as follows, with the number of brackets in parentheses: bulkhead no. 1 (one); no. 2 (three); no. 3 (two); no. 4 (five); no. 5 (five); no. 6 (three); no. 7 (five); no. 8 (seven); no. 9 (six); no. 10 (six); no. 11 (three); no. 12 (zero). The number of brackets per bulkhead varies, yet a certain pattern of use of brackets is definitive, as at least one bracket must be used for each set of three planks with carvel seams. On the bulkhead profile, the strake(s) that comprise the four carvel sets show either one bracket or two brackets per set. The brackets thus can be seen as fastenings for joining a set of strakes to the bulkheads rather than just as fastenings for individual planks. The brackets are recessed into at least two bulkhead planks (in some cases three bulkhead planks). Their lengths vary. Either shorter or longer brackets are attached over the lower and upper bulkheads in such a way that each bulkhead plank from the bottom to the top can be fixed by one of the brackets. Thus, the brackets can be also seen as a method for fastening a set of bulkhead planks.

The brackets can sustain tension in only one direction and do little to secure the bulkheads and planking in any other direction. The brackets, however, are used together with nails. The nails through the inner planking to the transverse components (bulkheads and frames) are not as strong as the brackets, but they strongly resist distortion in other directions. The use of the two different types of fastenings together creates a strong system to secure a plank shell and transverse component. Whether iron nails were driven from the innermost planking into the bulkheads when these components were assembled together is still arguable. The outer planking for sheathing obscures the identification of the use of iron nails in the inner planking. Part of the topmost plank that remains at bulkhead no. 7 exposes the outside surface of the innermost plank, where iron nails appear to have been driven originally from the outside of the inner planking to the bulkheads. They are identified as corrosion remains, and original nail holes are positioned where the bulkhead plank was originally placed. Although the original bulkhead plank is missing, its position can be ascertained by the location of an opening from the iron bracket. The use of Portland cement sealing a modern nail was observed. This nail is located where it can fasten the planking and frames together. A fastening method used in the Shinan ship demonstrates two transverse components: both frames and bulkheads are fastened to the hull planking by iron nails. Use of the same method of fastening is also taken into consideration in the Quanzhou ship.

Frames

Half frames are fitted against the bulkheads on the side toward midships: in other words, the frames on the six forward bulkheads are on the aft faces of the bulkheads, and aft of midships the frames are on the forward faces. There is no frame associated with bulkhead no. 12. Most of the frames observed in the current hull remains are not original, according to communication with staff of the Museum of Overseas Communication History during the 2009 investigation. Only the frames associated with bulkhead no. 1 are likely to be original, and these frame timbers are significantly deteriorated, but they are half frames (figure 4.17). The only evidence regarding the dimensions of the original frames comes from

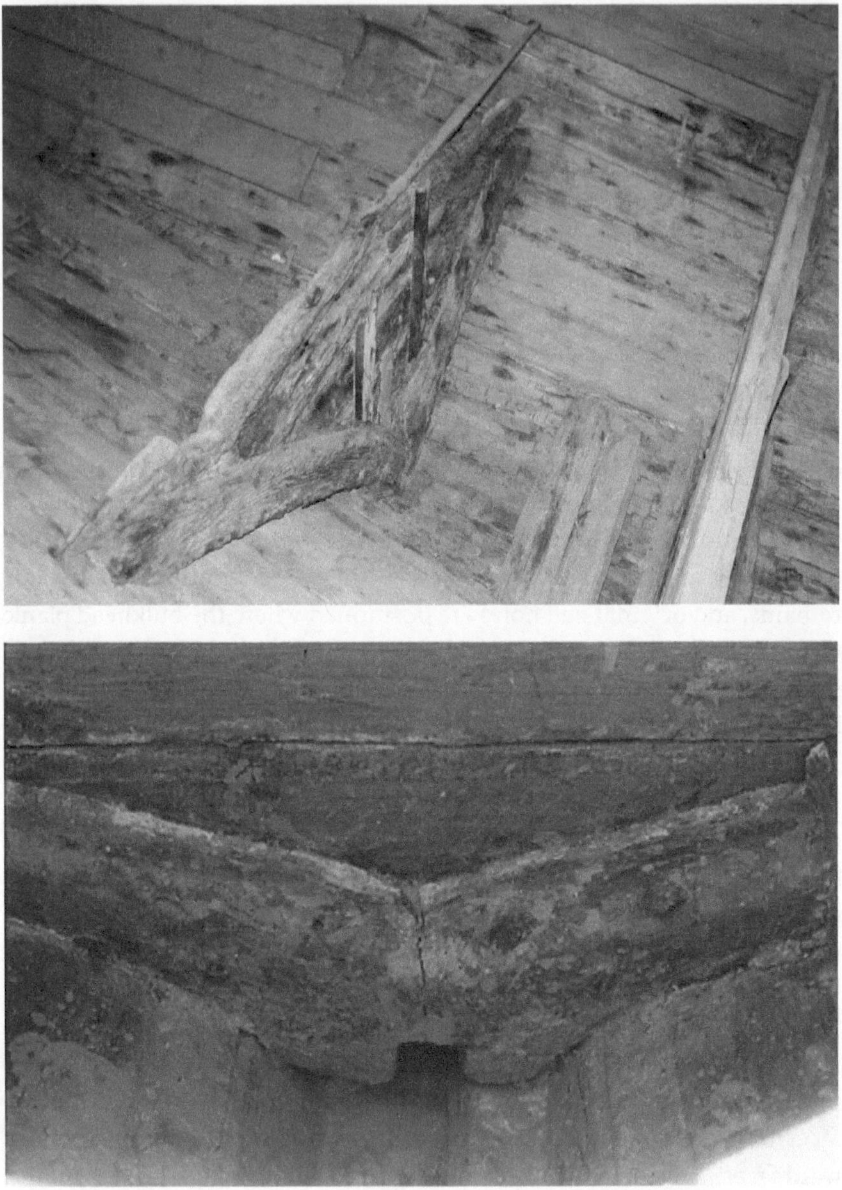

Figure 4.17. Top view of bulkhead no. 1, which has original half frames (*top*). Large half frames were attached to the bulkheads (*bottom*). (New photo [top] by author; older photo [bottom] courtesy of Museum of Overseas Communication History)

the surface of the bulkheads, where the outline of the upper edge of the original frames can be faintly seen, and from the planking, where the edge away from the bulkhead can be discerned in places. The faint lines indicate that the frame was approximately 280 millimeters at its widest, with sides of approximately 200 millimeters indicated by lines on the planking. These are fairly large dimensions for frames, which can be seen as auxiliary to the bulkheads. They were presumably intended to secure the bulkheads against deformation and against sliding forward and aft with the brackets. An old photograph shows the structure of the frames, consisting of two pieces of timber per side (figure 4.17). The timbers of the frames meet at the keel, where limbers are cut in the frames.

Other Hull Components

Two mast steps remain in the hull. Their dimensions are available in the published report (Museum of Overseas Communication History 1987: 22). A forward mast step is positioned on the forward side of the first bulkhead (figure 4.18). It measures 1.76 meters athwartships, 0.50 meters fore and aft, and 0.36 meters in depth in the middle. Mortises, measuring 240 × 210 millimeters and spaced about 400 millimeters apart, presumably indicate the position of two tabernacle cheeks. A mainmast step is located on the forward side of bulkhead no. 6 (figure 4.18). It is shaped to fit directly on the keel and hull planking out to the fifth strake. It measures 2.74 meters athwartships, 0.56 meters fore and aft, and 0.48 meters in depth in the middle. Two tabernacle mortises are placed about 375 millimeters apart and measure 240 × 210 millimeters. The mast steps are not now accessible for examination of their details, and how they were fastened to the hull has not been determined. Photographs taken during excavation show that there were substantial timbers running forward from the mainmast step to the frame on bulkhead no. 5 to secure the step against moving forward. The diagram in the report shows that the bottom of the forward mast step has a somewhat rounded shape where it fits into the bilge, whereas the bottom of the mainmast step has a flat shape fitted to the keel (Museum of Overseas Communication History 1987). A large notch on bulkhead no. 5 is thought to be related to the mast, which could possibly have been lowered aft in its tabernacle with the heel of the mast swinging forward through the gap in bulkhead no. 5 (Burningham and Green 1997; Green et al. 1998).

Figure 4.18. Photographs of the mainmast step (*above*) and forward mast step (*right*) after excavation. (Courtesy of Museum of Overseas Communication History)

There is an unidentified timber (recorded in an old photograph) placed on the current hull (figure 4.19). This timber is rectangular in shape, with a length of 3.00 meters and measuring 0.32 × 0.20 meters in cross section. It has a square hole (110 × 120 millimeters) at one end, and the other end is tenoned. According to the original report, this timber was discovered

Figure 4.19. *Top*: Timber discovered near the stern; its purpose has not been determined. *Bottom*: A cylinder-shaped timber, possibly part of a windlass. (Courtesy of Museum of Overseas Communication History)

outside the stern and is not related to the hull structure, but that presumption has not been verified.

A capstan or perhaps a windlass mentioned in the report is not visible in the current hull as displayed at the Quanzhou Ancient Shipwreck Gallery. Researchers found a cylinder-shaped timber with a length of 1.4 meters and a diameter of 0.35 meters. It has four notches in the midsection, and on one side are holes 130 millimeters in diameter. This is probably a windlass barrel, with holes for the pins or hand-spikes. Because of the loss of the actual object, a photograph is the only record available for this object (figure 4.19). One photograph indicates that this possible windlass might have been in the stern when it was discovered (see figure 4.2).

The excavation report also describes the discovery of the wooden post already mentioned, 3.00 meters long and with a rectangular cross section of 0.32 × 0.20 meters. Besides this solid timber, another four pieces of timber, an oar, and long bamboo poles were listed. However, identifying where these were located in the hull is not possible. The information about these additional findings was drawn from the Chinese reports and resources including photographs and videos in the Quanzhou Maritime Museum's custody.

Perspectives on Hull Construction

More precise dating of the Quanzhou ship and a closer examination of construction methods have yielded additional insights.

Hull Construction Date by Radiocarbon Dating

The initial determination of the date of the Quanzhou ship was based on study of its cargo. I attempted to collect new data regarding the hull construction date through radiocarbon dating of timber from the Quanzhou ship. Despite the fact that the amount of cargo found represents about 1 percent of the original load, the assemblage of the cargo remains indicates that the ship probably dates to the last quarter of the thirteenth century. In particular, the era of the minting of the coins (1265–74) was determined to be the latest date, after which the ship was sunk or abandoned. Radiocarbon dating of timbers provides a better understanding of the time of the ship's construction. Two pieces of timber from the hull were provided by the Museum of Overseas Communication History as specimens for

Table 4.1. Radiocarbon dates of timber samples from the Quanzhou ship

ID	Conventional radiocarbon age (yr. BP±1σ)	Calibrated age	
		1σ calibration date range	2σ calibration date range
Sample 1	685±20	AD 1280 (61.5%) AD 1298–1373 (6.7%) AD 1377	AD 1275 (70.1%) AD 1305–1363 (25.3%) AD 1385
Frame	810±20	AD 1220 (68.2%) AD 1254	AD 1186 (95.4%) AD 1269

detailed study during the inspection of the hull in 2008. The details of the timbers are not known, except for the fact that they derived from the Quanzhou ship. One of the specimens appears to be a portion of timber used for planking, perhaps from among the outermost planks. The wood species of these specimens do not conflict with the presumption that the specimen identified as Sample 1 is a part of a plank and the sample identified as Frame is a part of a frame (see chapter 6).

The two samples (that is, Sample 1 and Frame) were tested at the accelerator mass spectrometry (AMS) facility at the Paleo Labo Company in Japan. Table 4.1 shows the radiocarbon dating results. The conventional radiocarbon ages for the two specimens show approximately a hundred-year difference, with the Frame sample having an older date than that for Sample 1. The date variance between the two specimens may be explained in terms of the timeline of installation of the hull components. Hull frames are primary inner structures, while the outer planks of the Quanzhou ship are likely to have been assembled later than the frame. The construction of the Quanzhou ship thus could have begun earlier than the 1270s.

Construction of the Plank Shell

Not until the 1990s was a series of surveys performed involving detailed analysis of the hull construction method; teams from Western Australia conducted these surveys, which suggest that the Quanzhou ship is likely to have had plank-first construction. A primary aspect of bulkhead-first construction (in which bulkheads should be seen as solid frames) is that the bulkheads must be strongly fastened to the keel by large nails (or bolts). The use of nails was not confirmed during my inspection of the Quanzhou ship, as direct access to the hull was limited. However, there seem to have been no nails used to fasten the bulkhead planks to the keel. Indeed, the possibility that such large nails could have been driven into the thin bulkhead planks of the Quanzhou ship is unlikely.

Plank-first construction is inferred by looking at some innovations in the construction of the ship's plank shell. The rabbets of the planking that are arranged the same way, for example, could have facilitated the work of shipwrights assembling the inner plank shell. The projecting lips on the lower part of upper planks that appear on the outside are likewise helpful for plank-first construction. Through this method, shipwrights could have more easily joined the upper strake into the lower strake along the curvature, probably by fixing the rabbeted seam of the strakes in the midsection first, then pushing and fastening the upper strake from there gradually forward or aft.

Another innovation is the construction of the plank shell by fastening together a set of strakes. Each set is formed by strakes that are edge-joined by carvel seams, and the strake sets are fastened into the bulkheads with iron brackets. The distribution pattern of the brackets is concentrated in positions adjacent to the butts of the strakes. Because the butts are weak points of the plank shell, this way of positioning the brackets around butts to secure them follows a logical and deliberate construction principle. Furthermore, there are many brackets in strakes above clinker seams and fewer in strakes below clinker seams. As joints of the carvel-built strake sets, the clinker seams needed to be secured as important joint parts. The structural principle results from the shipwrights' understanding that the plank shell was the most fundamental structure for constructing the sturdy hull of the Quanzhou ship.

5

The Shinan Shipwreck

The Shinan ship discovered in Korea is another representative seagoing ship that originated in the East China Sea region. More commonly known as the Shinan shipwreck, it is dated to the first quarter of the fourteenth century. The importance of the Shinan shipwreck has been recognized since its discovery. Experts at the National Research Institute of Maritime Cultural Heritage (National Maritime Museum), located in Mokpo city in Korea's South Jeolla Province, have analyzed its structure and hydrodynamic and hydrostatic performance. Official resources available in Korean include excavation reports and analysis of the hull (Choi 2004; Kim 2004; Lee 2004; National Maritime Museum of Korea 2006a). The ship has also been discussed in English-language sources (Keith 1980; Keith and Buys 1981; Green 1983b; Green and Kim 1989; Lee 1991). I inspected the hull of the Shinan shipwreck in 2007 and 2009 with support from researchers at the National Research Institute. The following sections review the hull structure based on the previous research and assess some nonstructural elements, including historical background and cargo. The hull construction method is described in detail with new perspectives.

Site Background and the Ship's Discovery

The Shinan ship is an East Asian trader sunk around the early fourteenth century. Its hull is as large as that of the Quanzhou ship. The background of this shipwreck involves maritime trade between Korea, China, and Japan during the Yuan dynasty (AD 1271–1368). One of the bronze plumb weights recovered from the cargo bears the inscription Qingyuan Lu (慶元路, now known as Ningbo), which suggests that the ship sailed from that Chinese port. One of the discovered wooden strips or labels shows the name of the Japanese temple Tofukuji (東福寺), located in Kyoto, which suggests that the destination was Japan. This wooden strip or tag provides a date of 1322, described as "Zhizhi San Nian" (至治三年), dating to the Yuan dynasty, and the date range of bronze coins from the ship also supports the idea that the ship was engaged in trade in the early fourteenth century. The Yuan dynasty, considered as descended from the Mongol Empire and also regarded as an imperial dynasty of China, took over maritime hegemony from the Southern Song dynasty (AD 1127–1279) in 1279. The Yuan rulers adopted a strict policy to control overseas trade. In addition, the Mongol invasion produced a temporary decline in overseas diplomatic exchange, including communications between Japan and China. Under these circumstances, maritime trade persisted sporadically because of overseas demand for commodities and continued to benefit Japanese society, contributing much to Japan's material culture. The Shinan ship carried some antique and used tea wares on board (Mori 2008), which are testimony to the market demand.

Historical study and cargo analysis suggest that the Shinan ship can be classified as an East China Sea trader in the category "temple-shrine chartered ship" (Murai 2005). Temples and shrines in Japan authorized by the Japanese central rulers were allowed to invest in maritime trade to finance their building expenses. Chartered ships carried cargoes partly or entirely owned by a temple or shrine. They sailed overseas as authorized merchant ships and were particularly active from the thirteenth to the fifteenth century. Japanese temples and shrines were delegating the ships, but the owners of the chartered ships and the sailors could be Chinese.

The Shinan shipwreck was located about four kilometers offshore from Imjado Island, which is one of the islands of the Shinan-gun, in the Jeollanam-do archipelago (figure 5.1). The site is in a strait between two islands. The water depth is twenty meters, and the area is known as a narrows, difficult to navigate, where a tidal race flows at an average of 2.5 knots. The strong tide and the seabed sediments consisting of fine sands and gravels cause turbid water, which restricts underwater visibility to 150–200 millimeters. Seasonal excavations from mid-July to mid-November, the most favorable season, were conducted for nine years to complete the recovery of artifacts and the hull.

According to a site report, the site was initially discovered in 1975 by a local fisherman, who recovered numerous ceramics; immediately after this discovery, the site began to be disturbed by locals, creating concern among government officials (National Maritime Museum of Korea 2006b). An archaeological survey and excavation were conducted in

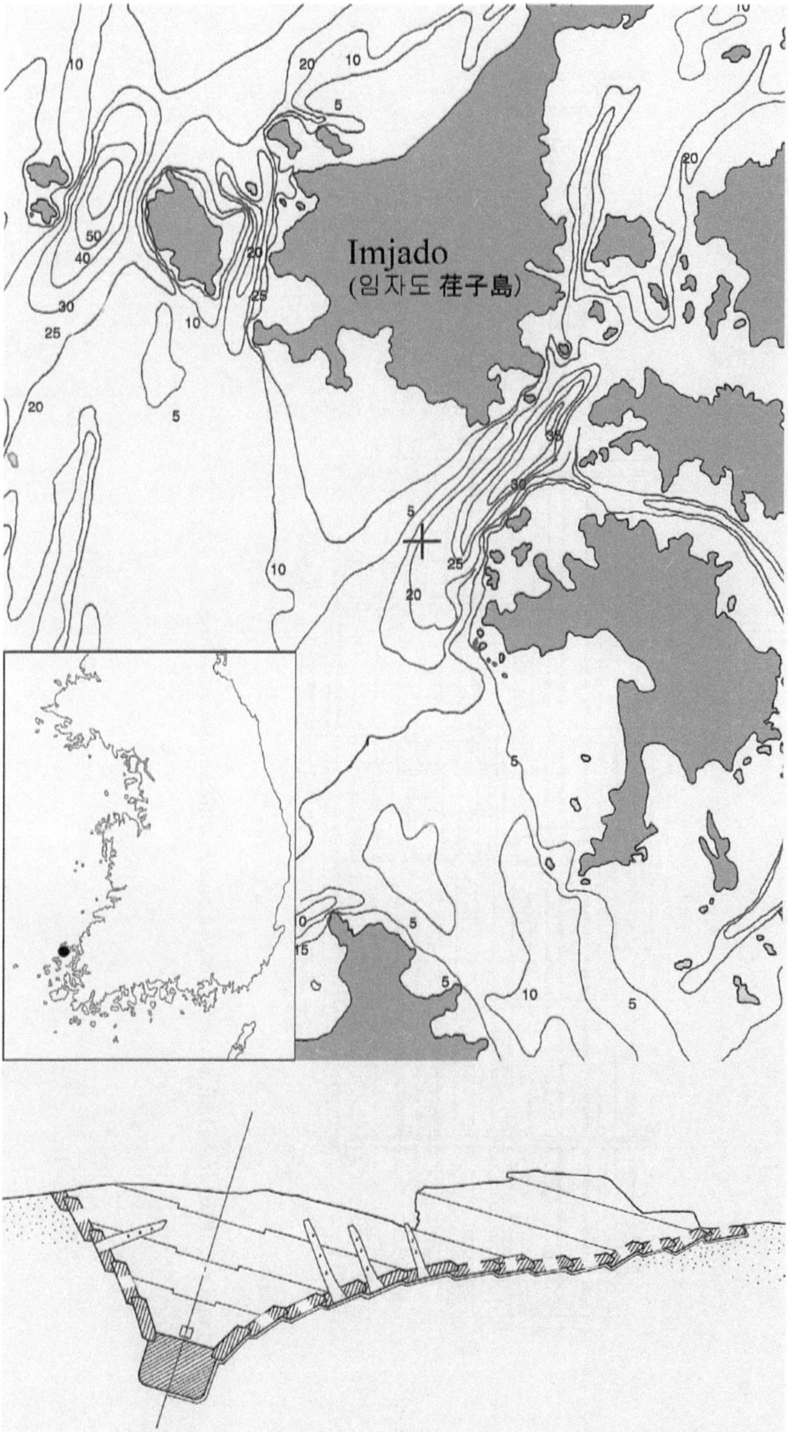

Figure 5.1. *Top*: The location of the Shinan shipwreck in the waters of the Shinan archipelago. *Bottom*: Cross section showing the buried condition of the Shinan shipwreck. (After, from top: National Maritime Museum of Korea 2006b; courtesy of National Maritime Museum of Korea)

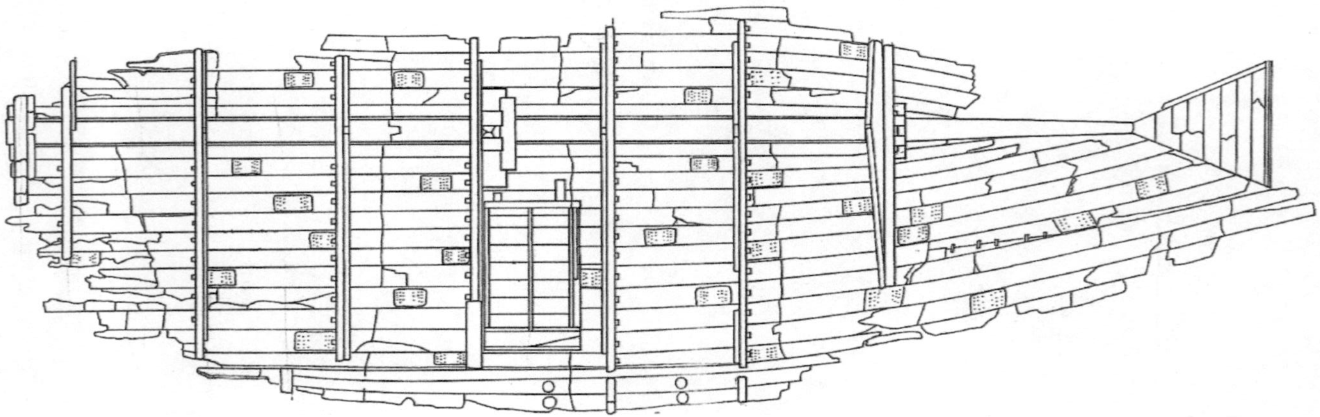

Figure 5.2. Plan of the hull. (Courtesy of National Maritime Museum of Korea)

collaboration with the Korean Navy Support Force from 1976 to 1984. In 1976 the Ministry of Cultural Properties conducted a site assessment survey to determine the significance of the site. Eleven full excavations were implemented during the following years. In the fifth and sixth underwater excavations, in 1979 and 1980, Korean survey teams identified the main structure of the hull, including the keel and well-preserved planking to the fourteenth strake on the starboard side and four strakes on the port side. These surveys found that the ship remains lay on the seabed inclining about 15 degrees to starboard (figure 5.1). The hull components were raised from the seabed in 1981 and 1983 by cooperating navy divers. Because of the difficult environmental conditions, the hull was dismantled, being cut into conveniently sized pieces underwater. Approximately 720 pieces were thereby recovered by the end of the work. The recovered parts consist of 479 portions of the hull's structural timbers and 223 portions of thin wooden planks that were sacrificial sheathing planks.

The Conservation Laboratory (which later became the National Research Institute of Maritime Cultural Heritage), established in Mokpo, conducted polyethylene glycol (PEG) conservation on the recovered hull remains from 1982 to 1999. In the same period, to facilitate interpretation of the structure of the ship, a scale model of the ship (completed in 1986) was constructed at 1:5 scale. After completion of the conservation treatment, the physical reconstruction of the Shinan shipwreck commenced in 1994, based on the scale model; it was completed in 2002 (figure 5.2).

Hull Dimensions and Volume of the Ship

The hull measures 28.4 meters in length and 6.6 meters in breadth. The extant starboard side includes a portion of the deck structure, which allowed Korean scholars to reconstruct the hull's original breadth and depth to the level of the deck with some confidence. According to a previous study, the reconstructed length of the original Shinan ship hull is 33.5 meters, its beam is 10.2 meters, and the vessel's depth is 3.6 meters, with a likely laden draft of about 2.6 meters (Choi 2004). This reconstruction posited that the freeboard of the Shinan ship was approximately 1 meter. Based on the reconstruction and posited freeboard, the displacement of the Shinan ship was calculated at 260.5 metric tons in that study. The light displacement of the ship, consisting of all its permanent equipment and hull weight, was calculated at 120 metric tons. Deadweight (the total weight of cargo, crew, and other necessary items associated with the persons on board) was also calculated. At 140 metric tons, the deadweight is slightly greater than the ship's light displacement.

Another study reconstructed the depth of the Shinan ship as approximately 3.75 meters and posited a laden draft of 3.1 meters, allowing only 0.65 meters of freeboard (Kim 2004). These figures are largely based on hypothetical load-line and freeboard. The estimated cargo carrying capacity of the Shinan ship thus would have been significantly less than that of the Quanzhou ship, for which capacity was calculated at more than 200 metric tons, even though the two ships show similar-sized hulls. This casts doubt on the estimation of the cargo carrying capacity of the Quanzhou ship; Chinese scholars may have assumed too much depth in the Quanzhou ship's hold.

Cargo Stowage

The assemblage of cargo characterizes the Shinan ship as a trader. Distribution of the artifacts in the remains of the hull as recorded during the excavations provides important information about the loading pattern of the cargo. The assemblage of artifacts raised during the 1977–84 excavations includes 20,661 pieces of ceramics, in addition to other major items consisting of 28,018 kilograms of coins, 1,017 pieces of wood, and 729 metal objects. Details about these artifacts are available in the official reports (National Maritime Museum of Korea 2006b). According to the site report, artifacts identified as highly valuable merchandise appear to have been in the midsection of the hull. The cargo assemblage is a range of commodities similar to that found with the Quanzhou ship. However, the quantity of cargo found with the Shinan ship is greater. The diversity of product quality reflects some specific market demand from both institutions and individuals. The site plan of the Shinan shipwreck (produced on a grid recording system) shows the distribution of the excavated cargo (figure 5.3). Some elements of the Shinan shipwreck cargo were discovered in wooden containers when the hull was excavated. Although most of these containers had been broken, the remains of ten containers showed the intact configuration with the ceramics packed inside. The ship carried a large load of wood (figure 5.3), identified as rosewood, *Dalbergia latifolia*. The wood was cut in lengths of approximately two meters and was stowed at the bottom of the hull. In addition, many copper coins were

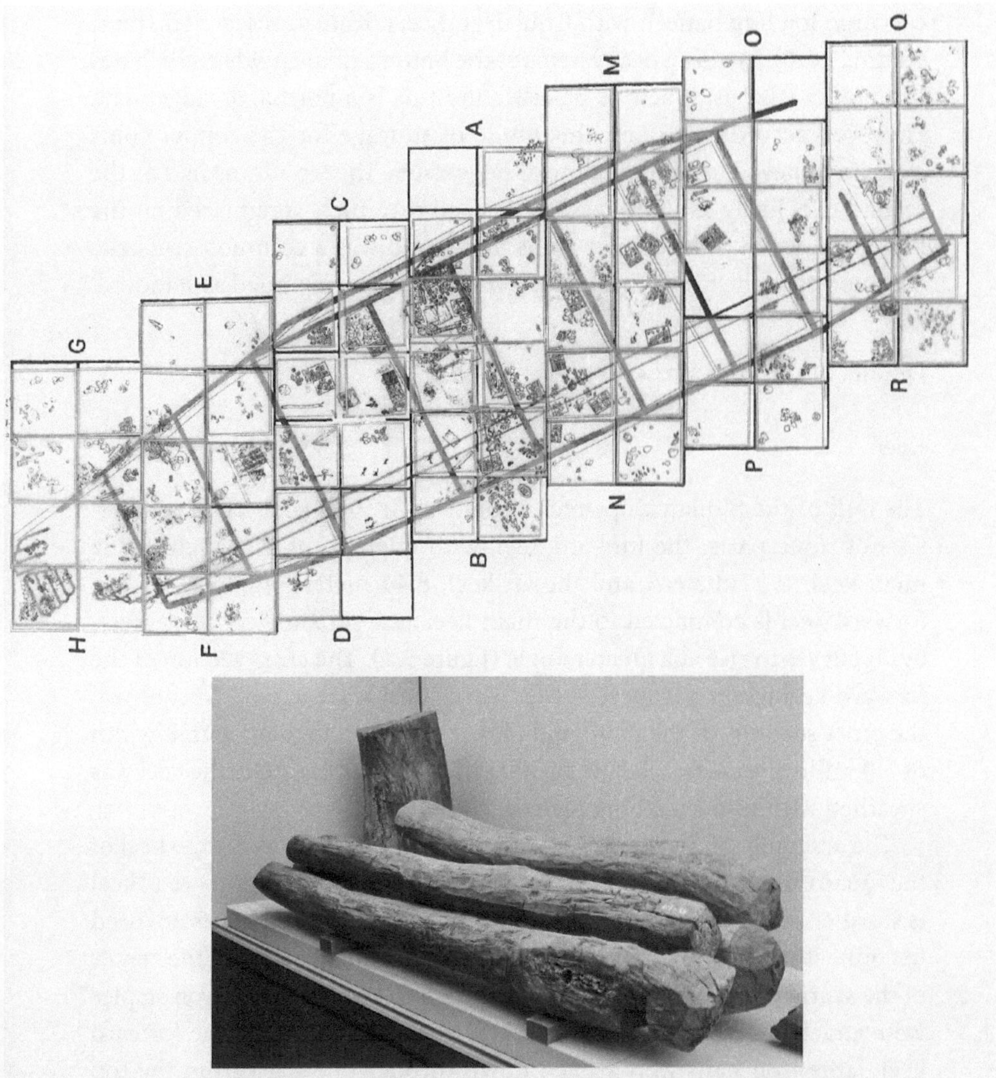

Figure 5.3. Diagram of the excavation using a grid system to record cargo location (*top*). Wood was a major cargo and could be stored in the hull to function as dunnage (*bottom*). (After, from top: National Maritime Museum of Korea 2006b; photo by author)

in the bottom part of the hull and placed on the keel. I discussed the vertical cargo loading pattern with Chul-Han Lee, a Korean researcher (pers. comm., 2010). Loading heavy items at the bottom of the holds contributes to stability (that is, it acts as ballast), and this is a normal arrangement. However, according to Lee, this mode of stowage for the copper coins appears to have blocked the flow of bilgewater. The wood was laid as the second tier, likely as dunnage. The packed ceramics were placed on the top of the wood. Like copper coins, rosewood was a common seaborne commodity at the time. This loading pattern can be regarded as standard.

Details of the Hull Structure

Keel

The hull of the Shinan shipwreck is built up from a keel. The keel consists of three parts: the forward keel, with a length of 6.79 meters; the main keel, 11.27 meters; and the aft keel, 8.44 meters (figure 5.4). The forward keel is connected to the main keel at approximately 10 degrees, but it curves to rise at a greater angle (figure 5.4). The cross section of the forward keel forms a trapezoid effective against wave action. In contrast, the cross sections of the main and aft keels are rectangular, with a width of 710 millimeters and a height of 500 millimeters. Originally the keel was sheathed with thin sheathing planks.

The assembly of the keel's three members is the same as for the keel of the Quanzhou ship. However, the structure of the Shinan shipwreck keel is sturdier than that of the Quanzhou ship, in that hooked scarfs are used in connecting the members (figure 5.4). There is an opening in the center of the scarfs where a pin was driven through to tighten them. Iron staples were attached on the sides of the scarfs to secure them. On the forward keel, large iron nails were driven down through the scarf from the top surface of the forward keel scarf into the main keel. The scarf join of the aft and main keels is also secured by a mortise and tenon, preventing the scarf from opening horizontally.

The Chinese shipbuilding custom of emplacing coins and a mirror, known as *bao shou kong*, was evident on the keel scarf; one mirror hole was recessed on the vertical face of the main keel's forward scarf, and an associated bronze mirror was identified (National Maritime Museum of Korea 2004). There are nine coin holes, in which seven copper coins

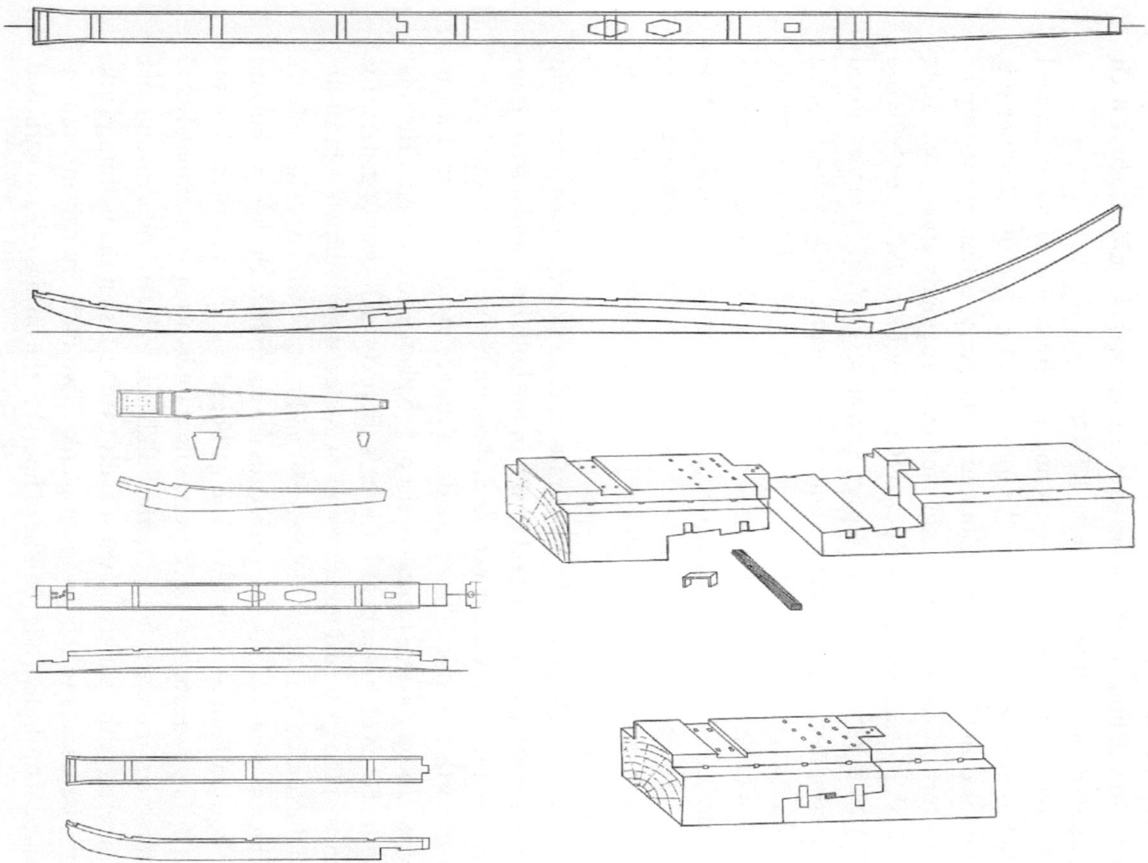

Figure 5.4. Keel members and a joined keel that clearly shows a hog and hooked scarfs of the keel join, secured by a wooden pin and iron clamps. (Courtesy of National Maritime Museum of Korea)

were found bearing the inscription Taiping Tongbao (太平通寶), manufactured during the reign of the second emperor (AD 976–83) of the Northern Song dynasty (AD 960–1179). These coins were placed on the horizontal face of the scarf of the main keel to the aft keel. Because there are iron nails driven down through the aft scarf, the horizontal distribution of the coins on the scarf may have restricted the options for driving nails through this scarf, a deficit partly compensated for by the mortise and tenon. Lozenge-shaped timbers were inlaid on the main keel's upper surface to seal holes in which were placed seeds; this could also have been a custom to bring good fortune to the ship.

As figure 5.4 shows, the keel is hogged; the center of the main keel is 200 millimeters higher than the fore and aft ends of it. Whether this feature was deliberately incorporated in the construction of the ship or is a distortion resulting from other forces, particularly gravity operating on the shipwreck, has been disputed (Green and Kim 1989; Lee 1991; Lee 2004). Most recently, in a final report, Korean researchers have mentioned the possibility that this hogging could be a deliberate design feature aimed at shifting the hull's buoyancy and weight-carrying aft (National Maritime Museum of Korea 2004). (The issue of whether the hogging was deliberate is revisited later this chapter.) According to the site report, despite the main keel being waterlogged, when it was recovered it still showed strong buoyancy—unlike other waterlogged timbers, including the forward and aft keels.

The keel is shaped to fit bulkheads and garboards. There are eight transverse recesses on the keel to slot in the bulkheads and a stern transom (one recess on the forward keel, three on the main keel, and four on the aft keel). The edges of the top sides of the keel have rabbets for the garboards. The garboard strakes consist of two planks on each side. The total length of each of the garboard strakes is approximately twenty meters. The planks of the garboard measure approximately 400 millimeters in width, and the thickness is approximately 180 millimeters, which is almost twice as thick as other planks. The garboards are cut to fit to the keel rabbet and the bottom bulkhead timbers. The garboard strakes are fastened to the keel by large square-sectioned iron nails at intervals of approximately 200 millimeters and are fastened to the bulkheads by a few square iron nails. Of interest is the pattern of those nails used to fasten the bottom planks of the bulkheads to the garboard planks. Two nails are driven from the surface of the bulkheads (bulkhead floors) to the insides of each garboard,

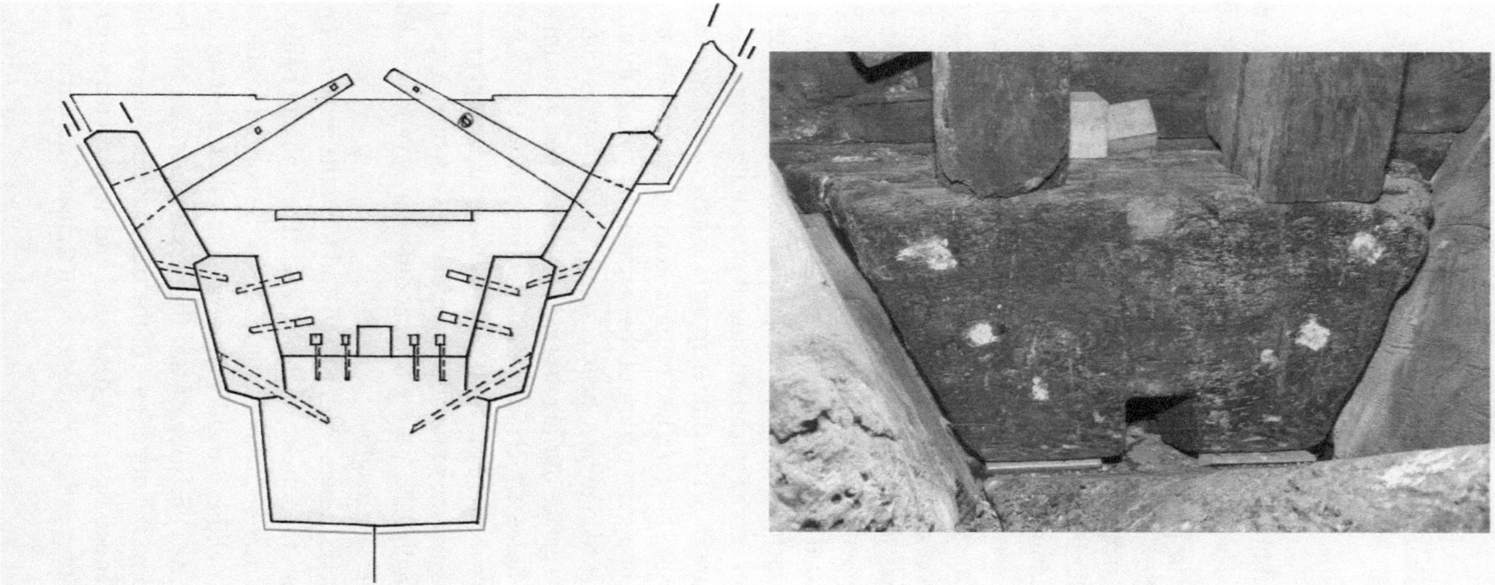

Figure 5.5. Diagram showing nails driven into the keel and garboards, and photograph of the bottom bulkhead plank showing the driven nails. (Courtesy of National Maritime Museum of Korea; photo by author)

whereas a larger nail is diagonally driven from the outer surface of the garboard into the keel (figure 5.5).

Structure of the Bow and Stern

The bow of the ship has a trapezoidal transom on the forward end of the forward keel. The angle of its rake as reconstructed appears to be approximately 60 degrees, but the original angle is open to question. The remains of the bow transom consist of two layers of planks: six outer planks and six inner planks (figure 5.6). The inner planks are attached to the thicker outer planks and are assembled to overlap the seams of the outer planks. The inner planks measure approximately 110 millimeters in thickness on average. They are joined by rabbeted carvel seams, whereas the outer planks are joined without rabbets (figure 5.6). The edge of the outer planks is recessed where the strakes meet the bow transom. The details of how the strake ends fitted to the transom have not been clearly explained. The upper part and bottom part of the bow transom are missing, but the original bow is estimated in an earlier study to have been built up to seven meters in height. A large baulk of timber is placed at the interior of the bow on the starboard side; it perpendicularly crosses over eight strakes (figure 5.7). This timber appears to have been part of a frame, and a similar timber is presumed to have been symmetrically fitted on the port side. Although a portion of the starboard planks is missing, the component can be seen to have been carefully cut with steps to fit neatly against the inner surface of the rabbeted clinker strakes. The timber has four rectangular holes, which were probably used to install beams. The interior of the bow forward of bulkhead no. 1 has a large empty space that seems to need framing or some other reinforcing structure. A few thwart beams may originally have been placed at different elevations in the bow to provide transverse strength.

Three solid baulks remain in the stern and form the bottom of a stern transom placed on the aft end of the aft keel. The transom baulks, with a consistent thickness of 220 millimeters from the bottom to the top, are thicker than the bulkhead planks. The transom rakes aft at an angle of 15 degrees. A frame consisting of two timbers is placed on the forward side of the transom. The upper part of the transom does not remain, but it could have extended up to the deck level and supported a superstructure including a stern gallery. Evidence for a transom-mounted rudder is

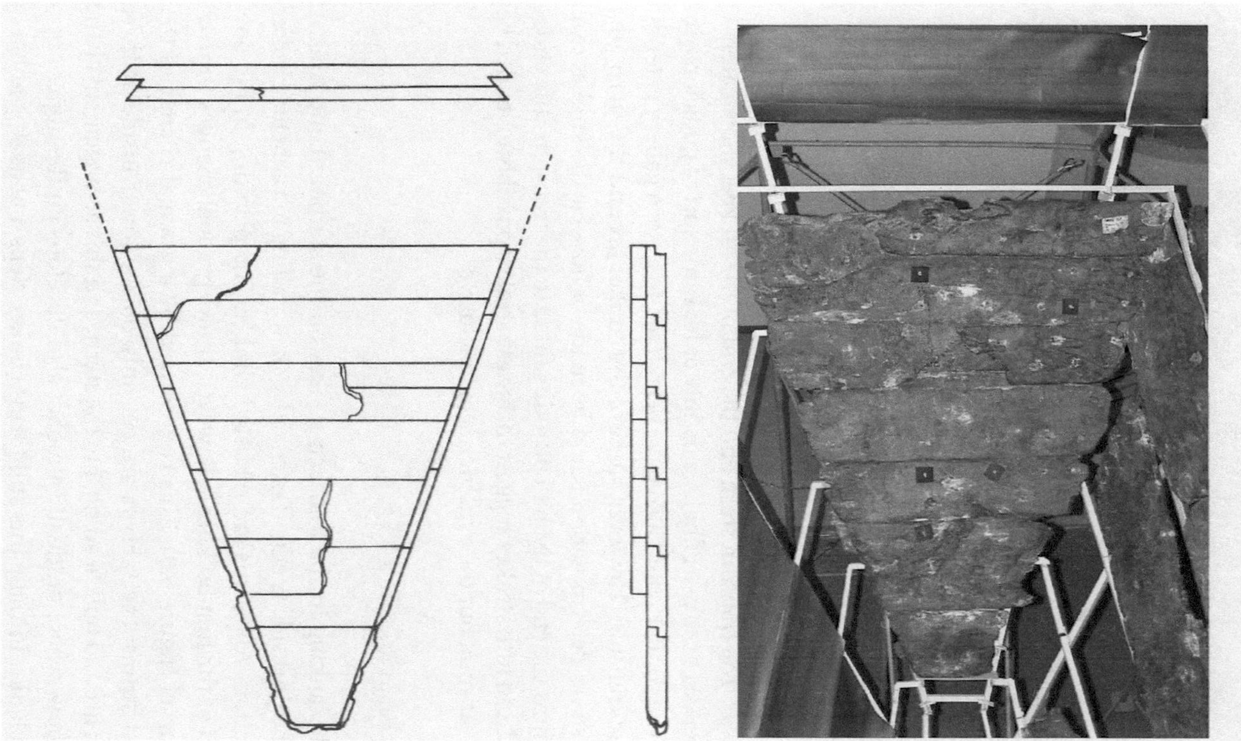

Figure 5.6. Drawing and photograph of the bow transom. (From National Maritime Museum of Korea 2004; photo by author)

Figure 5.7. Baulk on the inside at the starboard bow. (Photo by author)

unclear. The transom stern and the end of the aft keel show barely visible evidence of the use of nails, possibly to fix some sort of gudgeons or sockets. The rudder could also have been fixed to a stern gallery extending aft of the transom. However, most of the strakes around the stern are missing, so no determination could be made of how far the strakes extended and how the end of the hull planking fitted to the transom. The ends of the third and fifth strakes suggest that these strakes possibly extended beyond the transom, but the evidence is inconclusive.

Hull Planking

Hull planking in the fourteen strakes on the starboard and four strakes on the port side is edge-fastened by skew nailing in rabbeted clinker construction. All seams are rabbeted clinker. The planking in the bow, forward of the first bulkhead, however, shows gradual changes from clinker to carvel (figure 5.8). This is produced by a gradual change of the rabbeted seams; the rabbets are cut only on the lower inside of the plank from around bulkhead no. 1 to the stern, and this arrangement changes in the bow, where gradually a rabbet also develops on the upper outside of the planks. Initially this still forms clinker steps, but as the rabbet on the

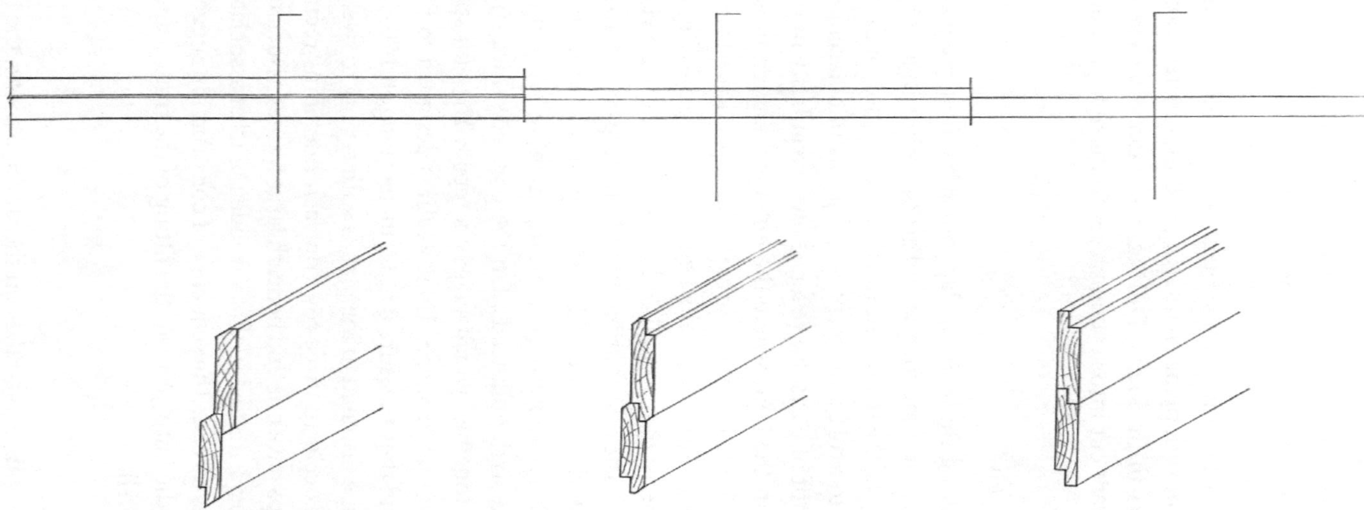

Figure 5.8. Hull planking arrangement in the fore part of the hull: rabbeted clinker changing to rabbeted carvel. (Courtesy of National Maritime Museum of Korea)

lower plank becomes deeper, it changes to a rabbeted carvel seam. The benefit of this arrangement has been explained as aligning the planks so that they meet the transom bow smoothly (Green and Kim 1989; National Maritime Museum of Korea 2004). Each strake comprises three or four planks, and the remains of the planks measure 450–500 millimeters in width (except for a bulwark plank measuring 600 millimeters) and from 110 to 120 millimeters in thickness. During the underwater excavation, most of the planks were cut into several pieces to make it possible to lift them (see figure 5.2).

Butt Joints and Butt Plates

The butts of the planking show simple lap joints, but in some places, including the garboards, there are tongue and groove joints (figure 5.9). The longitudinal assembly of the strakes shows that the location of the butts is carefully distributed to avoid alignment of the joints of adjacent strakes. The rule about the shift of butts, however, shows irregularity in the sixth and seventh strakes in the bow, where the butts are aligned in adjacent strakes. To secure the butts further, short lengths of plank are fastened on the inside to make what are usually called butt straps or plates (figure 5.9). No butts are placed beneath the bulkheads. The butt straps are visible on the interior of the hull. The distribution of the butt straps is approximately symmetrical on the starboard side and the four port side strakes remaining. However, there is an asymmetrical arrangement of butts and straps between the fifth and sixth bulkheads. On the port side third strake are two butt straps close together, which suggest a repair. The butt straps are rectangular or trapezoidal in shape. Their width is designed to fit to the width of the plank, and they measure 0.8–1.0 meters long and 70–100 millimeters thick. They are fastened to the planks with twenty to twenty-two nails, and these nails occur in three rows on each side of the rectangular butts. A few butt straps have one end beneath large frames associated with bulkheads, and one is positioned so that a wooden bracket associated with a bulkhead penetrates its end. The sequence of construction suggests that they were placed on the butts prior to the fitting of the frames and probably the brackets as well.

Bulwark Planks

Of the remains of the fourteen strakes on the starboard side, the two uppermost strakes are bulwarks. That is, the thirteenth strake is regarded

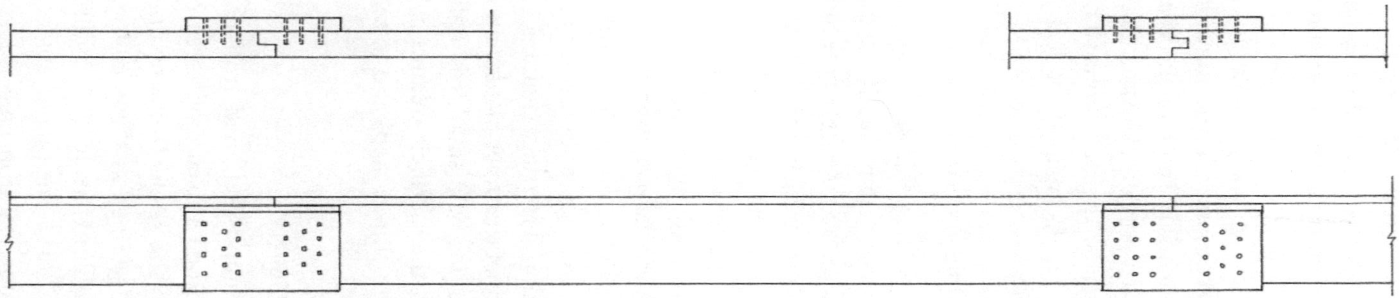

Figure 5.9. Butt joints (lap joint and tongue and groove joint) tightened by butt straps, also called butt plates. (Courtesy of National Maritime Museum of Korea)

as the first bulwark plank, and the fourteenth strake is regarded as the second bulwark plank. Details of the bulwarks have been disputed since the reassembly of the hull components. Four complete bulwark ports are visible on the remains of the bulwark planks. These bulwark ports are aligned in each bulwark strake to allow water to run away from the sides of the deck as scuppers (see glossary). The bulwarks are supported by bulwark stanchions. The remains of the stanchions extend from the second and third bulkheads, and a portion of a stanchion also remains between the fifth and sixth bulkheads. The stanchions sit on longitudinal bulwark planks. The longitudinal planks are in effect deck planks or may be a part of a coaming (figure 5.10).

Sacrificial Planks (Sheathing Planks)

The hull planking of the Shinan shipwreck consists of a single layer, but all the planks used to be sheathed by sacrificial planks. The thin wooden planks covered the outer surface of most of the hull planking and the keel to protect against attack by marine borers. The sacrificial planks make up approximately 30 percent of the recovered wooden remains associated with the hull of the Shinan shipwreck. When the hull was discovered, these sacrificial planks were still tightly fixed onto the outer surface of the innermost planks. The sacrificial planks were preserved with polyethylene glycol after recovery. They were not restored with the reconstructed hull and are still stored at the conservation center of the National Research Institute of Maritime Cultural Heritage (figure 5.10). The hull as exhibited exposes the outer surface of the inner planks, on which there is evidence of adhesive putty used along with small iron nails to affix the sacrificial planks. The thickness of the sacrificial planks is 10–20 millimeters, and their width varies to fit to clinker steps and to seal the seams of the hull planking.

Fastening of the Hull Planking

Square-sectioned and round-sectioned iron nails were used for hull planking (National Maritime Museum of Korea 2004: 54). They were diagonally driven through the seams from the upper planks to the lower planks. This is usually described as skew nailing. Intervals between the nails range from 150 millimeters to 200 millimeters (National Maritime Museum of Korea 2004). The planks are fastened to the bulkheads and their associated frames by iron nails. Most of these iron nails are

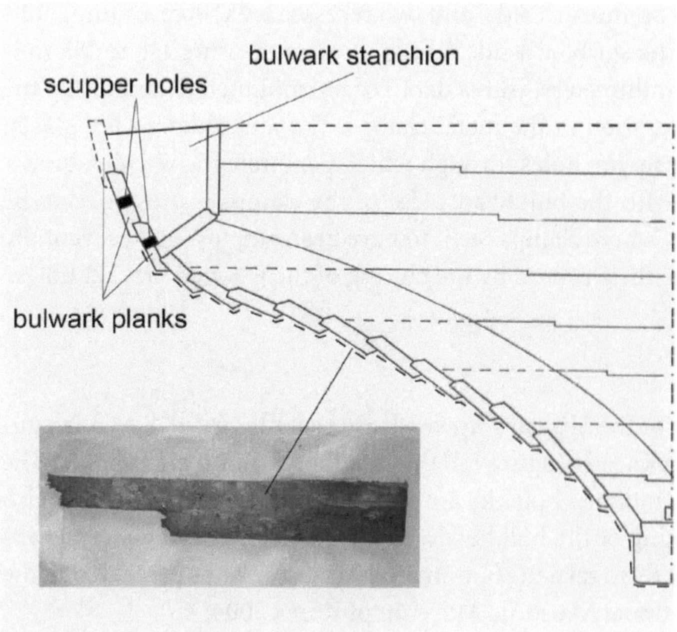

Figure 5.10. Drawing of the reconstructed coaming structure and photograph of a sheathing plank. (Courtesy of National Maritime Museum of Korea; photo by author)

considerably degraded, and they were replaced by modern metal nails when the hull was reassembled. These modern nails are said to have been driven into the original nail positions.

Although corrosion of the original nails was considerable, magnetite remnants of the original nails have remained in the planks. Yellow substances, probably sulfur compounds and other substances, surrounding the nail holes and modern nails are now visible in many places on the hull. The other corrosion observed on the outer surface of the planks is evident on the remnants of a few square metal clamps. The square clamps, measuring around 150 millimeters in length, were used to secure the scarf joint of the planks in the eighth strake on the starboard bow. This is the only scarf in the hull planking; all other plank joins are butts with lap joints or tongue and groove joints. The use of similar metal clamps is observed elsewhere on the outer surfaces of the planking; one square recess with a little metal remaining is visible near bulkhead no. 2 on the port side, one square recess with a little corroded metal remaining is visible around bulkhead no. 3 on the port side, and two recesses are visible around bulkhead no. 5 on the starboard side. These recesses measure 145 to 185 millimeters by 40 millimeters, with a depth of 180 millimeters, indicating the original configuration of the metal clamps. The metal clamps are placed adjacent to the square holes through which penetrate the wooden brackets associated with the bulkheads. Use of the clamps is presumed to be from one place, where clamps seem to have been employed to prevent the planks from splitting caused by the cutting of the wooden bracket holes.

Bulkheads

The bulkheads of the Shinan shipwreck are well preserved. Seven irregularly spaced bulkheads (figure 5.11) divide the hull into eight holds. The remains of the bulkhead planks are summarized in table 5.1, which also shows the spacing of the bulkheads, calculated based on the recesses indented on the keel to take the bottom bulkhead planks, as presented in the site report (National Maritime Museum of Korea 2004: 45).

Bulkhead no. 1 consists of thicker planks than the other bulkheads, and their thickness is consistent from the lower to the upper planks. Bulkheads nos. 2–7, however, have a different structure: the lowest planks of these bulkheads are thicker than the upper planks. These thick bulkhead planks have been described as "bulkhead floors" (Green and Kim 1989; National Maritime Museum of Korea 2004). The thickness of the bulkhead floors varies from 180 to 200 millimeters, compared with the upper

Table 5.1. Plank tiers and the spacing of the holds

Bulkhead	Remains of plank tiers	Hold	Spacing (m)
Bulkhead no. 1 (foremost)	6	1st hold (bow transom–BH no. 1)	5.01
Bulkhead no. 2	9	2nd hold (BH nos. 1–2)	2.70
Bulkhead no. 3	7	3rd hold (BH nos. 2–3)	2.58
Bulkhead no. 4	7	4th hold (BH nos. 3–4)	3.07
Bulkhead no. 5	7	5th hold (BH nos. 4–5)	2.54
Bulkhead no. 6	6	6th hold (BH nos. 5–6)	2.47
Bulkhead no. 7 (aftmost)	4	7th hold (BH nos. 6–7)	2.62
Transom	3	8th hold (BH no. 7–transom)	1.10

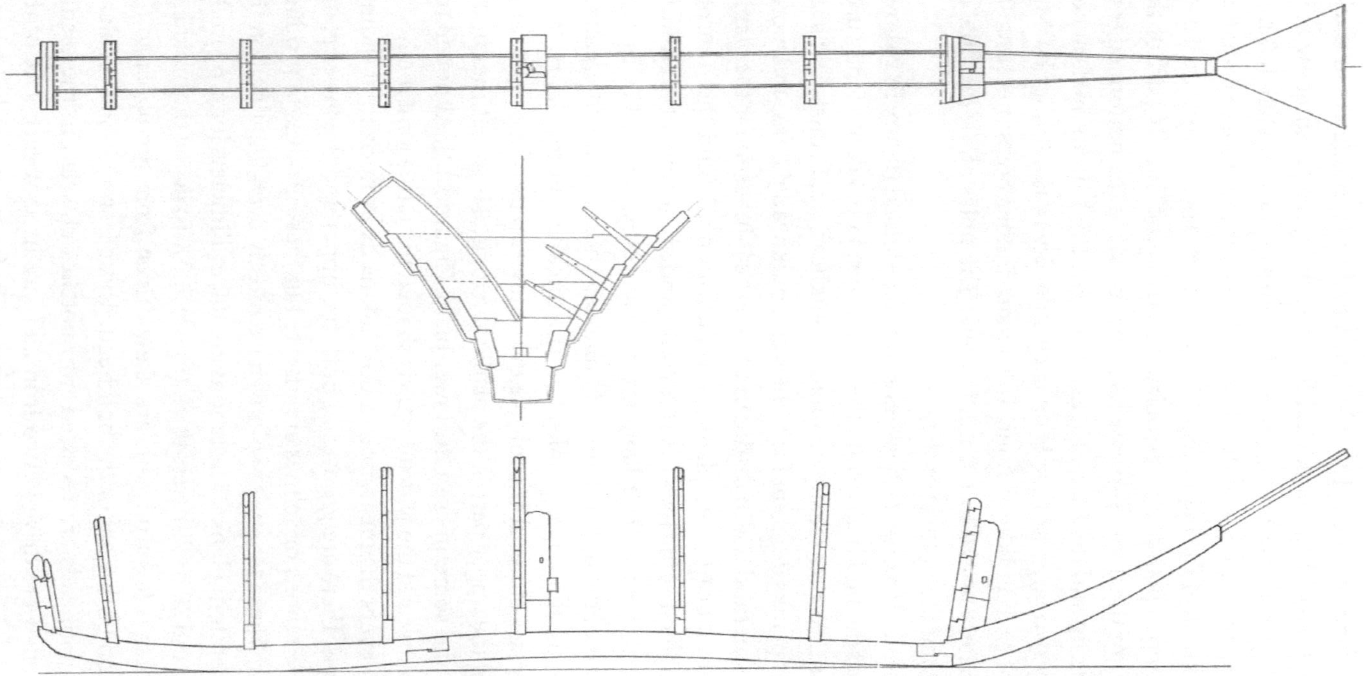

Figure 5.11. Profile showing the position of the bulkheads and cross section showing floors into which the end of the large half frame was slotted. (Courtesy of National Maritime Museum of Korea)

planks, measuring between 110 millimeters and 120 millimeters in thickness. The thickness of the bulkhead floors is designed to combine with the thinner upper planks and associated frames (approximately 100 millimeters thick) in such a way that the thickness of the floors is equal to the total thickness of the upper planks and frames. These floors are a base to hold the upper planks and associated frames; the middle part of the upper surface of the floors is partly recessed to slot in the bottom edge of the next plank up and the frame (figure 5.11). The thick floors are fastened to the keel by two large iron nails on each side. There is a limber hole on the bottom of each bulkhead floor. The square limber holes measure 100 to 110 millimeters by 120 to 130 millimeters. The holes allow the flow of bilgewater along the bottom of the hull.

The bulkhead planks up to deck level are edge-joined (figure 5.12 shows the remains of the bulkhead plank tiers). Bulkhead no. 1 consistently uses a half-lap joint. As previously mentioned, bulkhead no. 1 consists of a series of thicker planks different from the other bulkhead planks, and moreover the joint method is also distinctive from the joint used for the planks in the other bulkheads. The planks of bulkhead no. 1 (and the transom stern) are rabbeted at the lower part forward and at the upper part toward the stern, which forms a half-lap joint (see figures 5.11 and 5.12). In contrast, the seams of the upper planks of bulkheads nos. 2–7 are different: for these bulkheads, the upper planks, half the thickness of the floors, form ridges or recesses either at the middle of the lower part of the plank or at the middle of the upper part of the plank; the tongue is sometimes in the upper plank and sometimes in the lower plank (figure 5.12). Although the groove is shallow and the wide tongue is shaped to fit, this joint method is effective to prevent disintegration of the bulkheads by horizontal sliding. In some seams, there are grooves in both the upper and the lower plank edge. Multiple tenons or splines are inserted in these grooves to produce a strong joint. These tenons or splines have mostly been identified on the seams of the bottom planks. The tenons are 100 millimeters long, 60 millimeters wide, and 20 millimeters thick (figure 5.13). To fasten the bulkhead planks together, a few iron nails are skew driven from the upper to the lower plank on both sides of the bulkhead. These iron nails are regularly spaced. The nailheads are recessed into notches chiseled in the planks before the nails were driven. Two iron nails are used at each side of the floor, and a regular pattern—such as four nails for the second plank, six nails for the third, and eight nails for the fourth—is evident at each side

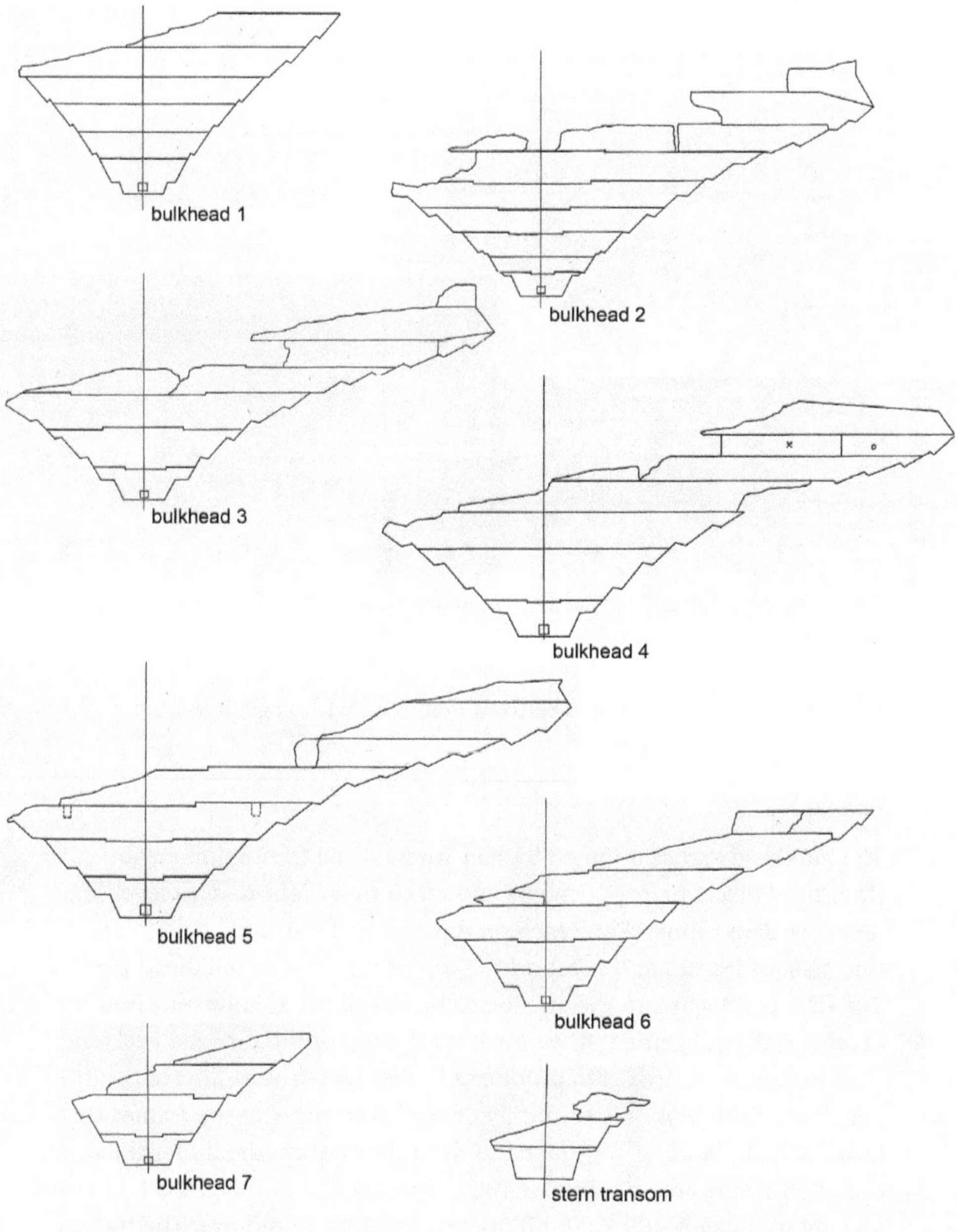

Figure 5.12. Drawing of bulkhead planks from bulkhead no. 1 (*top left*) to the stern transom (*bottom right*). (Courtesy of National Maritime Museum of Korea)

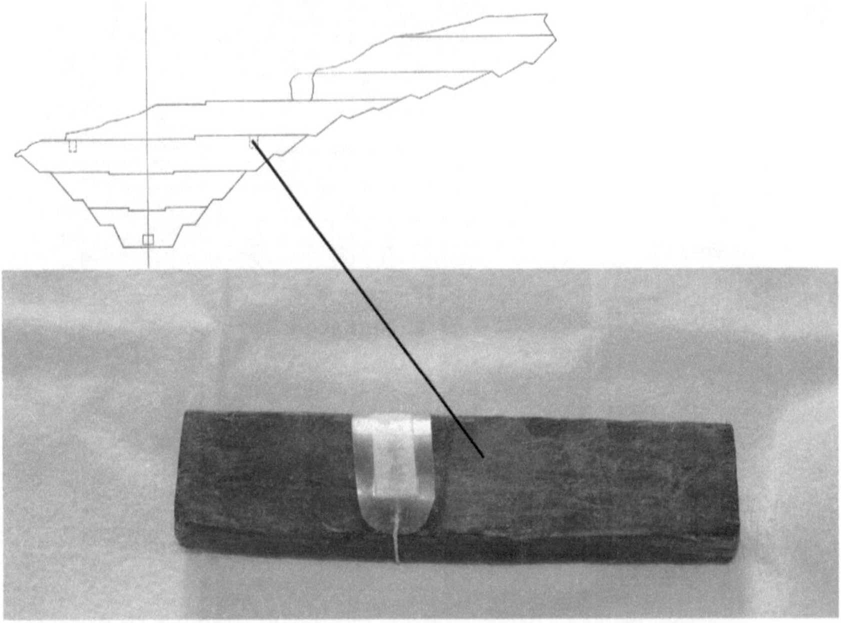

Figure 5.13. Tenon used for a joint of the bulkhead planks. (Photo by author)

of the plank. The bulkhead seams are caulked with vegetable fiber and a putty of tung oil and lime.

Frames

The bulkheads are reinforced by half frames. The frame timbers are cut from solid pieces of large timbers and reach from floor to deck level (figure 5.14). Two timbers (one each on the port and starboard sides) constitute a set of frames and meet on the top of the bottom bulkhead plank. The frames attached to the aft side of bulkhead no. 1, however, show an exceptional arrangement: they meet on the top of the forward keel, and their bottom ends are cut to produce a limber hole. The outline of the upper edge of the frames is smoothly curved. The edges of the frames that face the hull planking are shaped to fit to the clinker. The dimensions of each half frame are approximately 100 millimeters in thickness (siding) and approximately 300–500 millimeters in width (molding). The frames are fastened to the bulkheads by numerous iron nails that seem to be randomly driven (figure 5.14). One of the large half frames on one side has more than thirty nails fastening it to the bulkhead.

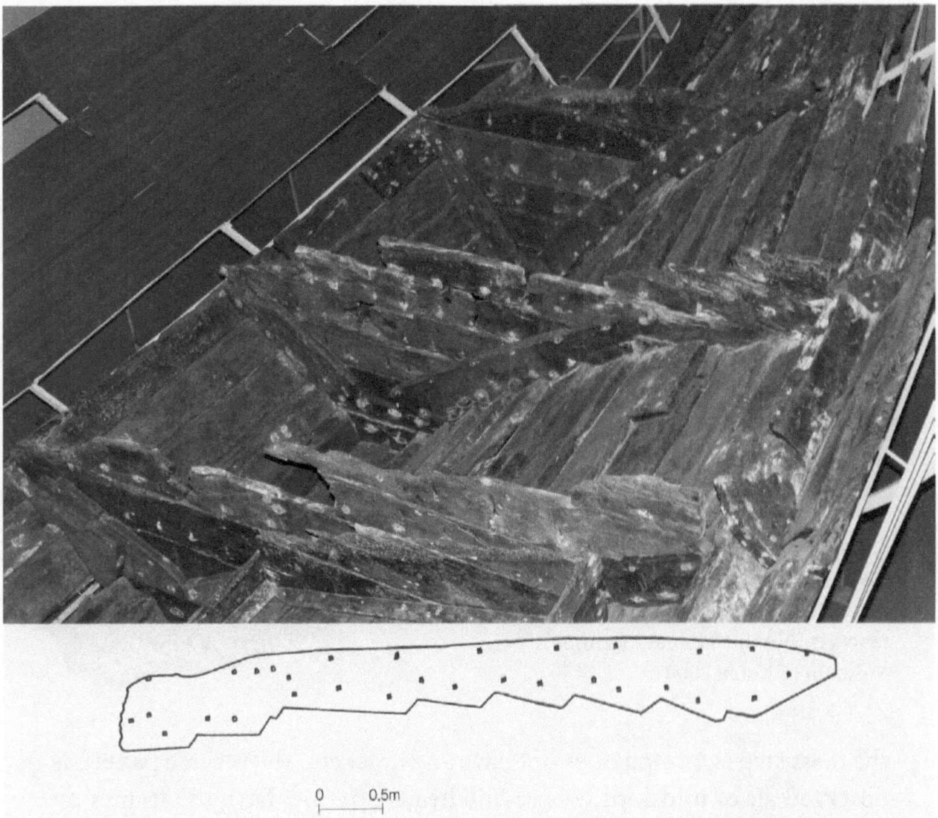

Figure 5.14. Half frames of the aft side of bulkheads 1–3 (*top*) and drawing of a half frame showing the irregular distribution of the nails (*bottom*). (After National Maritime Museum of Korea 2004; photo by author)

Wooden Brackets

The method used to fasten hull planking and bulkheads together on the Shinan shipwreck is similar to that used in the construction of the Quanzhou ship. The combination of frames and brackets on the bulkheads is a common technique. In the case of the Shinan shipwreck, instead of the iron brackets observed on the Quanzhou ship, tapered wooden brackets (which have been called "stiffeners") are used (Green and Kim 1989; National Maritime Museum of Korea 2004).

On the bulkheads forward of midships, which includes bulkheads nos. 2 and 3, brackets are used on the forward side and then frames on the aft side. Bulkhead no. 1 has the frames on the aft side, but the fore side where

Figure 5.15. Bird's-eye view from the aft side of the hull showing the arrangement of the frames and brackets. (From National Maritime Museum of Korea 2004)

the mast step is located does not have any brackets. The reverse pattern is observed aft of midships, where bulkheads nos. 4–7 have the frames on the forward side and brackets on the aft side (figure 5.15). These components effectively lock the bulkheads to prevent them from moving fore and aft. The brackets also tie the strakes to the bulkheads.

The wooden brackets are square at the bottom (outer) end and taper to the other end (figure 5.16). Length varies, but many brackets average 800 millimeters in length, with the bottom being approximately 100 millimeters square. Two examples were measured during this research: one is 950 millimeters long with the bottom being 114 millimeters square, and the other is 824 millimeters long with the bottom being 105 millimeters square (figure 5.16). Square holes in the strakes having the same dimensions as the bottom of the brackets allow the brackets to penetrate from the outside of the hull planking. Presumably the holes were made slightly smaller than the brackets so that the brackets would not pull right through the planking. The brackets are fastened to the bulkheads by iron nails. The number of iron nails ranges from three to five, depending on the length of the brackets.

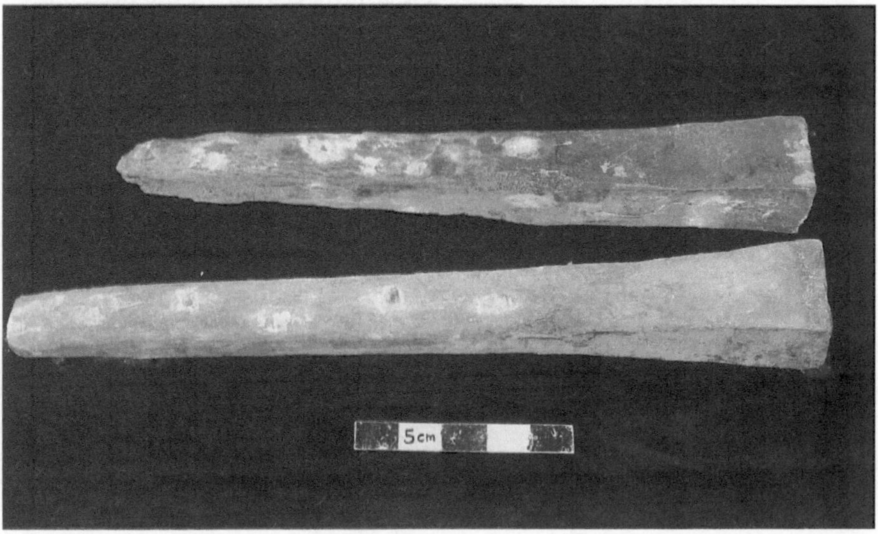

Figure 5.16. Wooden brackets fastened into the surface of the bulkhead planks (*above*) and wooden brackets shown for scale (*below*). (Photos by author)

Other Hull Components

The Shinan shipwreck yielded some mast detail and the remains of a tank believed to have been used for water or likely for cargo stowage.

Mast Steps

Mast steps and the heel of the mainmast remain on the hull of the Shinan shipwreck. A foremast step is identified forward of bulkhead no. 1 on the forward keel, and a mainmast step is placed on the forward side of bulkhead no. 4 (figure 5.17). Both mast steps have flat bottoms firmly placed on the keel, and their sides are shaped to fit to the clinker steps of the hull planking. A few iron nails are skew driven from the side of the mast step down to the keel and garboards for fastening. There are limber holes on the bottom of the mast step for bilgewater to flow through.

On each mast step are two tabernacle holes measuring 250 × 160 millimeters, with a depth of 150 millimeters, for tabernacle cheeks. The mainmast tabernacle cheeks remain and are about 500 millimeters apart, which shows the diameter of the heel of the mainmast. Three irregularly shaped timbers remain between the tabernacle cheeks on the mainmast step. These appear to be the heel of the mainmast. The timbers are fixed to the tabernacle cheeks by a locking pin, which penetrates across the three timbers (perhaps originally four timbers) and the tabernacle cheeks. A square-shaped timber is horizontally fixed to the forward side of the tabernacle cheeks, tying them together. Except for this transverse lock, the structural remains of the foremast tabernacle and mast heel are the same as those of the mainmast. The foremast step uses the same system, though only two timbers of the mast heel remain between the tabernacle cheeks.

The assembly of the mast heel indicates that the masts of the Shinan ship originally consisted of multiple timbers assembled to make one mast pole. On later Chinese ships, this arrangement was used with iron bands to hold the timbers together. The two mast steps show that the Shinan ship had at least two masts. Korean researchers have suggested that a mizzenmast might have been located on the upper part of the stern (National Maritime Museum of Korea 2004), but the possibility remains that Chinese merchant ships of the twelfth and thirteenth centuries had only two masts.

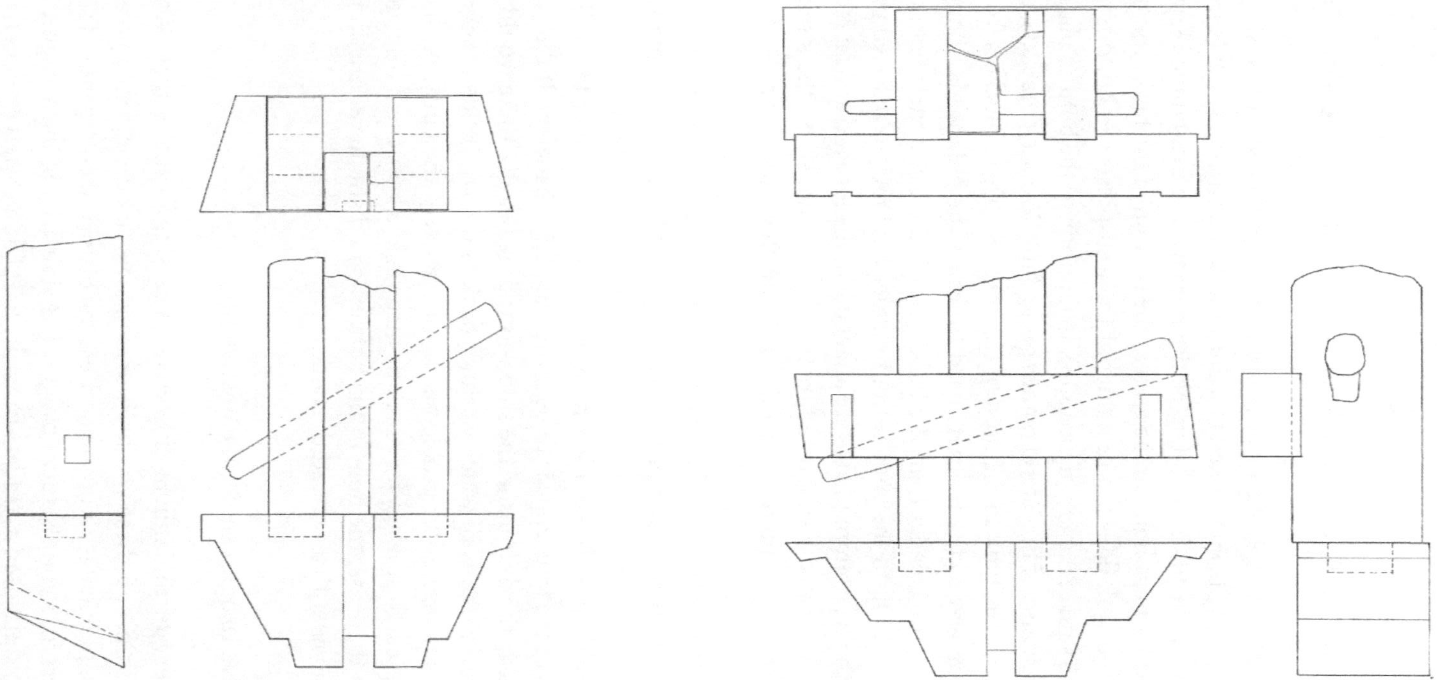

Figure 5.17. Drawings of the foremast step (*left*) and the mainmast step (*right*). (Courtesy of National Maritime Museum of Korea)

Tank

The remains of a tank were identified on the starboard side of the Shinan shipwreck. The wooden tank is placed in the fourth hold, between bulkheads nos. 3 and 4. According to a site report, the structure shows evidence of having been caulked to achieve watertightness, which is seen as indicative of use to carry water (National Maritime Museum of Korea 2004: 58). A drainage hole can be seen at the bottom of the tank. The upper part of the tank does not remain, but the well-preserved bottom has an area of 5.5 square meters (figure 5.18). In addition to the bottom planks are partial remains of the lateral planks of the tank. Wooden brackets fasten the sides of the tank to the bottom (figure 5.18). These brackets function the same way as the brackets used to fasten the bulkheads, though these brackets are smaller than the bulkhead brackets and their bottoms are dovetailed to fit to the tank bottom planks. Six brackets inserted into the ends of the six bottom planks are attached on the forward and aft faces of the tank. The tank is supported by two frames that sit on the strakes from the third strake up to the tenth strake on the starboard side of the hull. The frames are formed to fit to the clinker steps. Another tank is thought to have been placed on the port side. If they were used as water tanks, two tanks would have provided enough drinking water for the Shinan ship to be engaged in long-distance trade. There is, however, a considerable problem with the interpretation that they are water tanks. The dimensions of the tank appear to be more than 2.5 × 2 meters. If the tank was about two-thirds full and the ship was rolling heavily, there would be about five metric tons of water washing back and forth inside the tank. Such large tanks would be problematic. Alternatively, the tank may have been used as cargo stowage for precious materials to keep them from getting wet. In that case, the hole in the bottom could still have functioned for draining water that entered the tank.

Comparative Insights on Hull Construction

The discovery of the Shinan shipwreck is as significant as that of the Quanzhou ship, and these two hulls are valuable resources for comparative and complementary study to address the historical emergence of seagoing traders for long-distance voyages. The similarity in the background of these two ships used as traders from the East China Sea is evident from

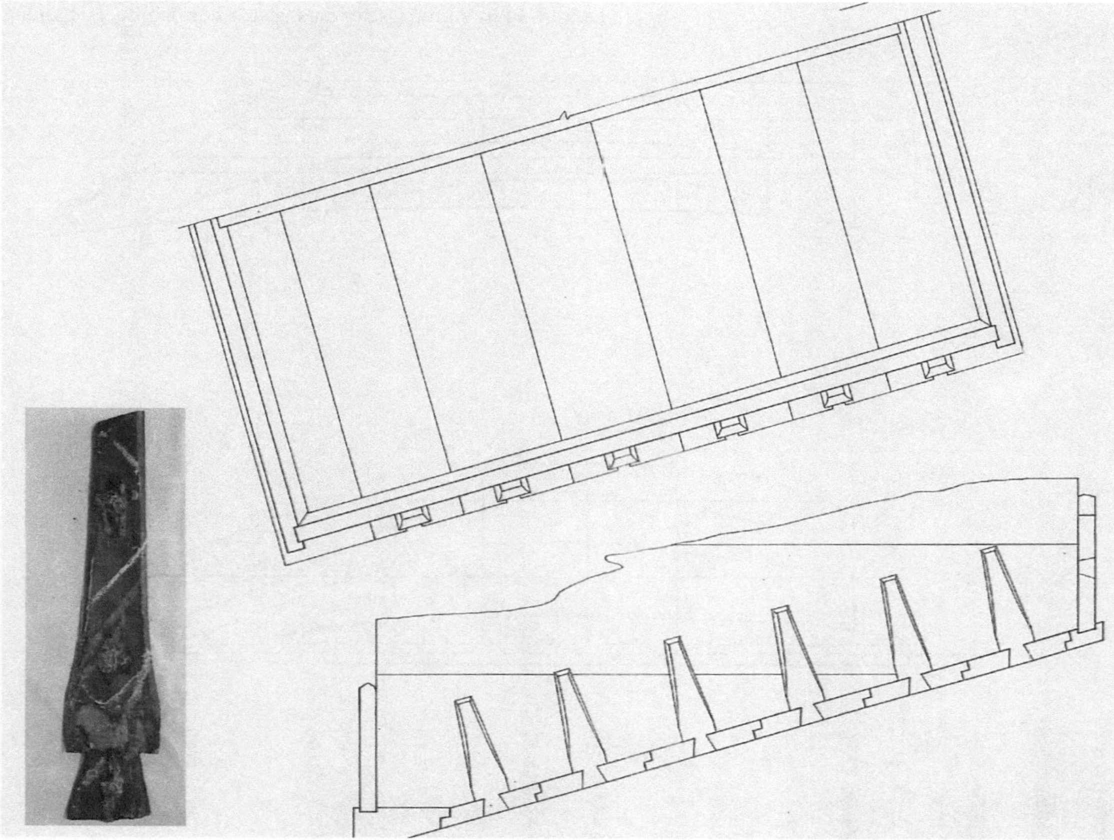

Figure 5.18. Drawings of the tank and photograph of a wooden bracket to fix the tank. (Courtesy of National Maritime Museum of Korea; photo by author)

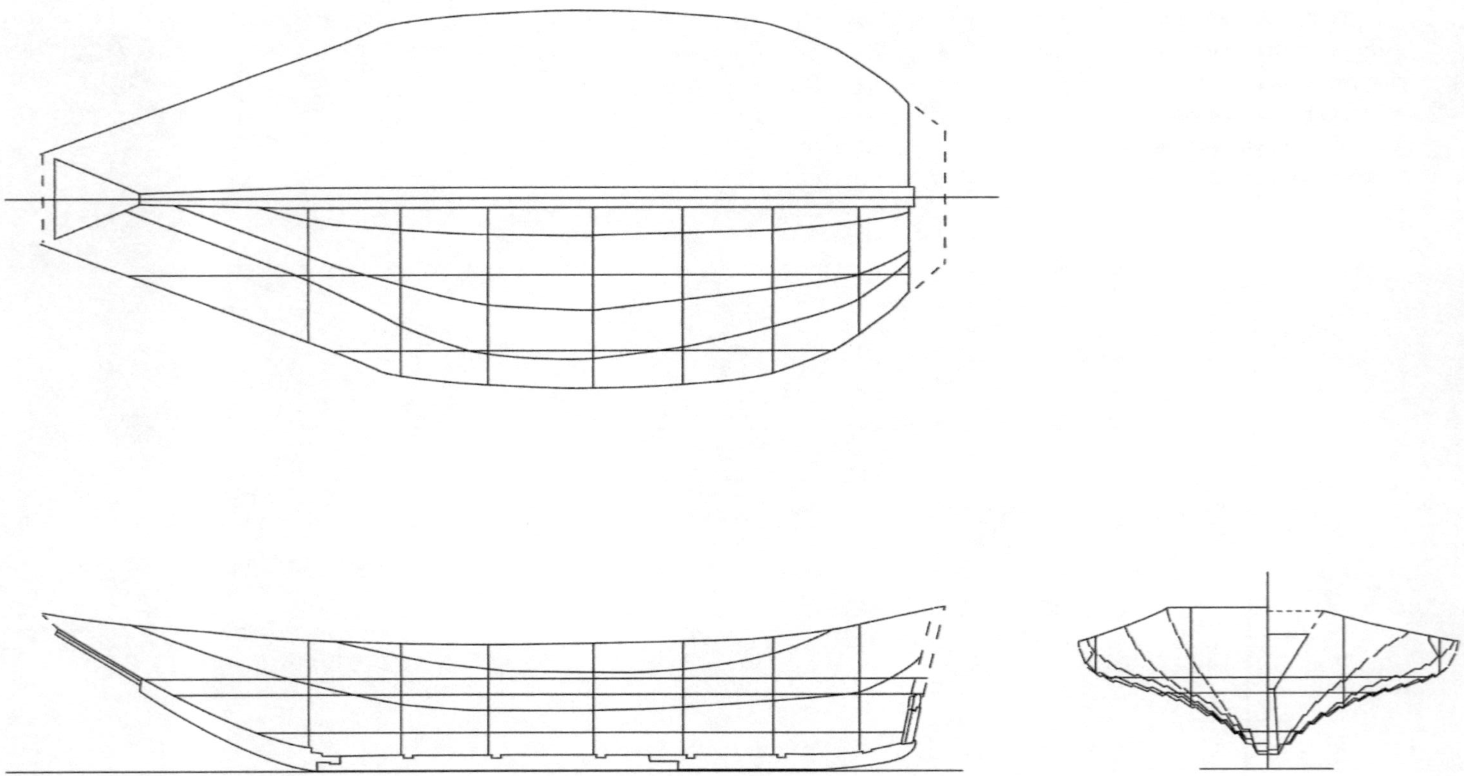

Figure 5.19. Shinan shipwreck, ship lines. (Produced by author)

their cargo assemblages as well as their hull structure and construction methods. Standardization in the construction of merchant ships is represented in the remains of these two ships: a keel consisting of three members, edge-joined hull planking, and bulkheads as transverse components.

The well-preserved hull allowed for the production of a plan of the ship lines for the Shinan shipwreck. The process of reconstructing the original ship lines revealed an idiosyncratic feature in the hull of the Shinan shipwreck: the hog of this ship's keel has prompted disagreement as to its uniqueness. In general, the hogging of a keel is regarded as a distortion. During the reconstruction of the Shinan shipwreck, the hull distortion had already been recognized. The ship lines plan indicates that the keel hog is not a deliberate design (figure 5.19). The deadrise of the hull shows a smooth transition on the presumption that the main keel was originally straight or even curved upward toward the bow and stern. If the ship had originally been designed with a hogged keel, the aft transom would have dragged deep in the water, which would have resulted in poor sailing performance.

Seen in comparative perspective, the ship lines plans of the Quanzhou ship and Shinan shipwreck allow us to understand the configuration of the two traders with hulls representative of the East China Sea shipbuilding tradition. They have in common a wide breadth and a shallow draft, with a V-shaped cross section indicative of effective resistance against leeway. The hull configuration reflected in the ship lines demonstrates these ships' seaworthiness, capable of carrying out long-distance voyages across the East China Sea.

Structural differences between the ships are evident too. The bulkheads, for example, are fewer in number in the Shinan shipwreck than in the Quanzhou ship, and the adequacy of transverse strength needs to be carefully considered. The Shinan ship bulkheads show an innovation in the structural integrity of the bulkhead planks to resist lateral distortion by forces imparted by mast and rigging. The pattern of nails driven into the bulkhead planks shows regularity, and mortises and tenons are used for the plank seams. The way of joining the bulkhead planks—by forming ridges or recesses at the middle—could have provided the integrity needed to resist such forces.

The use of bulkhead floors, thicker than the upper bulkhead planks, is remarked as a characteristic of the Shinan shipwreck's bulkheads, though some Southeast Asian shipwrecks dating to later periods use a similar structure. The size of the bulkhead floors provides space beside the limber holes for nails to be driven into the keel. In general, a large limber hole in the center of the bulkhead is likely to be an obstacle to nailing bulkhead planks to the keel. For this reason, a limber hole does not often appear on the center line; usually there are two small limbers, one on either side of the center line. Compared to the thin bulkhead planks of the Quanzhou ship, the large and thick floors of the Shinan shipwreck are regarded as facilitating the fastening of the bulkheads to the keel with nails. Iron nails are also used for fastening the floor to the garboards. The garboards are fastened to the keel by large nails driven from outside, but for the fastening of the floors the nails are skew driven from inside. The pattern of fastening shows that the garboards probably came before the bottom of the bulkheads. If the bulkheads were fitted first, the fastenings would be through the garboard from the outside into the bulkhead floor.

The hull of the Shinan shipwreck shows layered planking. The hull planks function as sheathing planks rather than serving other purposes, such as contributing to longitudinal strength. For example, the arrangement of the outer planks of the Shinan shipwreck is not designed to cover the seams of inner planks, like those of the Quanzhou ship. Furthermore, each strake of the Shinan shipwreck is individually fastened to the bulkheads by wooden brackets; they appear on every strake except the garboards. In contrast, the iron brackets in the Quanzhou ship serve sets of strakes. The technique of using brackets for individual strakes might result in considerable weakening of the hull, as all the square holes cut in the individual strakes align.

The underlying idea of using brackets to fasten the planking to the bulkheads is shared by the Quanzhou and Shinan ships, while the configuration and material of the brackets differ. As a result, the structural efficiency of the two ships is also considered to be different. The L-shaped iron brackets of the Quanzhou ship, firmly fastened into both the plank shell and the bulkheads, would have had good tensile strength, but the wooden brackets of the Shinan ship might have pulled through the planking under tensile load. Yet the wooden brackets would have probably been better for preventing fore and aft movement of the bulkheads; the iron brackets seem too thin to resist that kind of distortion.

6

Ship Construction Materials

Focusing on construction materials identified in the East China Sea shipbuilding tradition permits us to understand technological innovations in more detail. The hull component study extends to the examination of ship fastenings and timbers—the organs of ship structure, to be analyzed from an anatomical perspective. Recent investigations have widened the scope of comparative study by assessing the remains of iron nails from the Shinan shipwreck and timber samples from the Quanzhou ship. These specimens were examined in conjunction with data from the Takashima underwater site, a historical site associated with the thirteenth-century Mongol Empire's attempted invasion of Japan. A large fleet of the Yuan dynasty (AD 1271–1368), which was established in China by the fifth Mongolian emperor, was sent to Japan in 1274. However, the invasion attempt failed, as did a second in 1281, and during the second invasion in particular, many ships were lost. The loss of the fleet transporting Mongolian, Chinese, and Korean troops is evidenced by the discovery of diverse artifacts related to weaponry and personal items as well as by numerous timber fragments from ships. Many of the ship remains discovered had originated from the East China Sea regions, and they complement archaeological study of East Asian ships of the thirteenth and fourteenth centuries.

My main aim in the first part of this chapter is to identify the characteristics of seagoing ships that had historically been constructed along the Chinese coast of the East China Sea, where iron materials were adopted for fastenings. The assemblage of wood species represented in the hull timbers of excavated ships is another avenue for addressing regional characteristics in shipbuilding traditions. Available data on wood species assemblages from several excavated ships in East Asia are presented as an additional component of the characterization. Combining these aspects and focusing on construction materials also allows us to review the supply chains for iron fastenings and timbers during the construction period of the East China Sea ships.

The Takashima Underwater Site

The historical background of the site has been reviewed in an archaeological site report and more recently in an English-language volume (Agency for Cultural Affairs 2000; Delgado 2009). In the thirteenth century, Genghis Khan led the expansion of the Mongol Empire in the Eurasian continent. Success came in China through one of his descendants, Kublai Khan. Kublai founded the Yuan dynasty after the conquest of the northern part of China, and then he subjugated Korea using both Mongolian and Chinese forces. Failure of diplomatic approaches to a Japanese ruler made him determined to put Japan under his control through military power as well. The first attack on Japan took place in 1274, but the fleet and soldiers were withdrawn immediately after landing in northern Kyushu and destroying Hakata city. The incomplete first attack convinced Kublai of the necessity of organizing a greater force.

A Thirteenth-Century Battlefield

Seven years after the first invasion, the second invasion was implemented with the dispatch of larger fleets. By the time of the second invasion of Japan, Kublai could obtain naval forces from the Northern Song dynasty (AD 960–1127). Before the advent of the Yuan dynasty, the Northern Song dynasty had reigned over the southern Chinese coasts, where many ships were based and shipbuilding industries had been established, but the Song dynasty was finally conquered in 1279. After the successful campaign, Kublai decided to send invasion forces to Japan again. According to the dynasty chronicles *Yuan Shi* (History of Yuan, 元史) and *Shin Yuan Shi* (Editorial history of Yuan, 新元史), in the second invasion two separate fleets were organized from different regions. The first fleet consisted of about 900 ships carrying approximately 40,000 soldiers and sailed from Masan (now Mokpo) at the tip of the Korean Peninsula; the second fleet consisted of about 3,500 ships carrying approximately 100,000 soldiers and sailed from Ningbo, a historical port of middle and southern China. The number of ships and soldiers may be exaggerated in the Chinese sources. However, the fact that two fleets from different

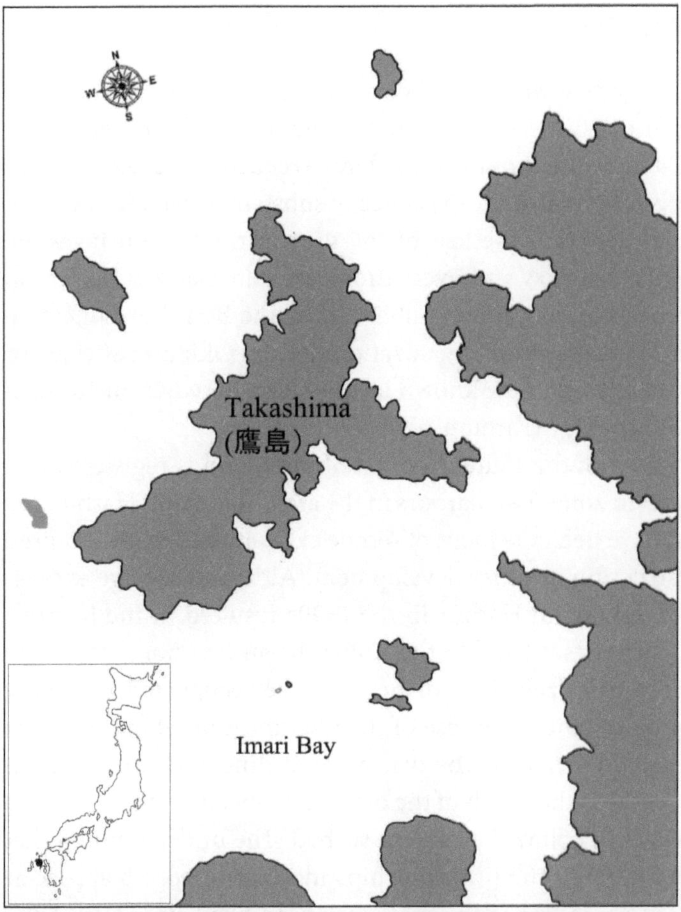

Map 6.1. Takashima Island in Imari Bay, Nagasaki prefecture.

points of origin left for Japanese waters is clearly mentioned in the historical texts. The two fleets initially intended to rendezvous in Japanese waters. After some disorder that caused delay of the missions, the two fleets were finally assembled together and anchored south of Takashima Island, located in present-day Imari Bay in the Nagasaki prefecture (map 6.1). The fleets concentrated in the bay suffered from resistance attacks by Japanese local clans, who were much more prepared than during the first invasion. In addition, a violent storm (later known among the Japanese people as the Divine Wind) struck the fleet in July 1281. Many ships sank to the bottom of the bay, and the rest of them could no longer fulfill the mission of conquest.

Archaeological Evidence

Imari Bay is a large bay with a surface area of 120 square kilometers, a maximum depth of 50 meters, and an average depth of 20 meters. Takashima is located at the mouth of the bay. Archaeological expeditions since the 1980s have revealed the existence of substantial remains beneath the seabed, resulting from the loss of the fleet in the second invasion. The archaeological surveys and excavations are summarized herein on the basis of archaeological reports published by the Board of Education of Takashima and the nonprofit organization Kyushu Okinawa Society for Underwater Archaeology, now known as the Asian Research Institute of Underwater Archaeology (Kimura 2006).

The area offshore of the southern coast of the island is registered as a nationally protected zone. Two harbors in the area, Tokonami Harbor and Kozaki Harbor, have been the focus of rescue excavations legally required preceding construction or other development. Although salvage excavation offshore of Tokonami Harbor in the 1990s resulted in the identification of many artifacts related to the Kublai Khan invasion, after 2000 the zone offshore of Kozaki Harbor began to be recognized as having greater excavation potential because of the identification of hull remains and better-preserved artifacts. The discovery of nine anchors in situ in 1994 and 1995 revealed the depth of the original seabed (at the time of the Mongolian invasion), below the current seabed. The understanding that the anchors had moored the invasion ships increased people's appreciation of this site as the setting of the shipwreck of Kublai Khan's fleet. During the rescue excavations in 2001 and 2002, 2,432 artifacts were recovered from the site. These excavations revealed that buried remains had been substantially disturbed by wave and tidal forces.

As noted, the historical records indicate that the fleet consisted of two units—one that departed from Ningbo and another that sailed from the Korean Peninsula. Analysis of the major types of ceramics discovered showed that most of the artifacts were of Chinese origin (Mori 2001), though some important Goryeo dynasty products were recovered from the waters, including sherds of Goryeo celadon wares and a Goryeo-style bronze statue. Study of the remains of timbers possibly associated with the hull structure likewise suggested that a large proportion of the remains exhibit characteristics of Chinese shipbuilding technology (Sasaki 2008, 2015).

Most of the remains gathered through the above-mentioned surveys and excavations were fragments, but a few can be confidently identified as important resources relevant to the study of shipbuilding technology. For example, two bulkhead planks recovered from the site offer more detail for interpretation than do the other disarticulated timbers recovered, and they are clearly associated with one of the ships constructed in the East China Sea region (figure 6.1). The bulkhead remains consist of two large planks; the upper edge of the larger bulkhead plank measures approximately 5.7 meters in length, and the upper seam of the smaller bulkhead plank joining the larger one measures approximately 4.5 meters in length. They were originally fastened by several iron nails, and the two timbers still held together when they were discovered. However, the degraded iron nails came apart during the excavation process. Some of the corroded nails were further examined, and the outcome of that study is reviewed later in this chapter.

As noted, the rescue excavation offshore of Kozaki Harbor in Takashima revealed nine large wooden anchors. They are another resource for detailed interpretation. All the anchors have an identical structure indicative of their Chinese origin. They appear to be the oldest wooden anchors ever found in the East Asian regions. While wooden anchors are neither hull components nor involved directly in a ship's construction, they were a critical element in a vessel's outfitting and have been reviewed extensively in previous studies of East Asian nautical technology (Needham 1971; Worcester 1971; Li 1998). The configuration of wooden anchors with stone stocks is important for addressing the historical development of East Asian anchors, which can be elucidated by pursuing transitions in the utilization of stocks (Kimura et al. 2011).

The preservation status of these anchors varies. They are all compound anchors consisting of a wooden shank and arms with separate stone stocks (table 6.1). The wooden anchors' shanks and arms are mostly broken; however, the components of anchors nos. 2, 3, and 4 remain relatively intact. Anchor no. 3 is the largest of the nine (figure 6.2). The remains of its shank are 2.74 meters in length, and the width of the midsection measures about 300 millimeters. One arm is in mostly intact condition, with a length of 3.15 meters. The shank is estimated to have originally been as long as 5 meters (Takashima-cho Board of Education 1996). The broken remnants of the other smaller anchors exhibit an average shank length of more than 2 meters. The remains of the shank of anchor no. 2 are 2.55

Figure 6.1. Two bulkhead planks recovered from the Takashima underwater site. (Drawings by author)

Table 6.1. Remains of the components of the composite anchors

	Shank	Arms	Loose tenon	Stakes (to hold stone stocks)	Stone stocks
Anchor no. 1	N/A	One arm	N/A	N/A	Identified
Anchor no. 2	Identified	Two arms	Identified	Identified	Identified
Anchor no. 3	Identified	Two arms	N/A	Identified	Identified
Anchor no. 4	Identified	Two arms	Identified	Identified	Identified
Anchor no. 5	N/A	Broken pieces	Broken pieces	Identified	Identified
Anchor no. 6	Broken pieces	Broken pieces	N/A	N/A	Identified
Anchor no. 7	Broken pieces	Broken pieces	Broken pieces	Identified	Identified
Anchor no. 8	N/A	N/A	N/A	N/A	Identified
Anchor no. 9	Broken pieces	Broken pieces	Broken pieces	N/A	Identified

Source: From Takashima-cho Board of Education 1996.

meters in length, and the midsection width is 170 millimeters. In anchor no. 4, the shank is 2.10 meters long, measuring 170 millimeters in width at the midsection.

The anchors are well enough preserved for us to infer their original configuration and construction methods. The shank of the anchor is a baulk of timber, and the tip of the shank is diamond-shaped where the two arms are attached. The timber of the arms is chamfered, and each arm is neatly fitted to the shank by a mortise-and-tenon joint. Besides the mortises and tenons, a piece of wood penetrating through the tip of the shank to the arms joins them together on the crown. Ropes made of bamboo secure the shank and arms. Some anchors have either one or two trapezoidal wooden plates that are attached on both surfaces of the anchor crown by iron nails. They provide additional structural strength to the junction of the shank and arms. Each anchor originally had two arms. The remains of some arms were toed with iron sheet metal to form flukes, but these have been heavily corroded. All anchors seem to have had two separate stones used as a set for an anchor stock (figure 6.2). The two stones are separately attached to the middle of the shank and are placed between a pair of wooden stakes that penetrate the shank through holes, the stones thus being fixed perpendicular to the arms and held in place with bamboo ropes.

Anchor stocks made of stone have been widely used on East Asian ships for a long time. Chinese ships, which began to be involved in international trading around the tenth century, were probably equipped with

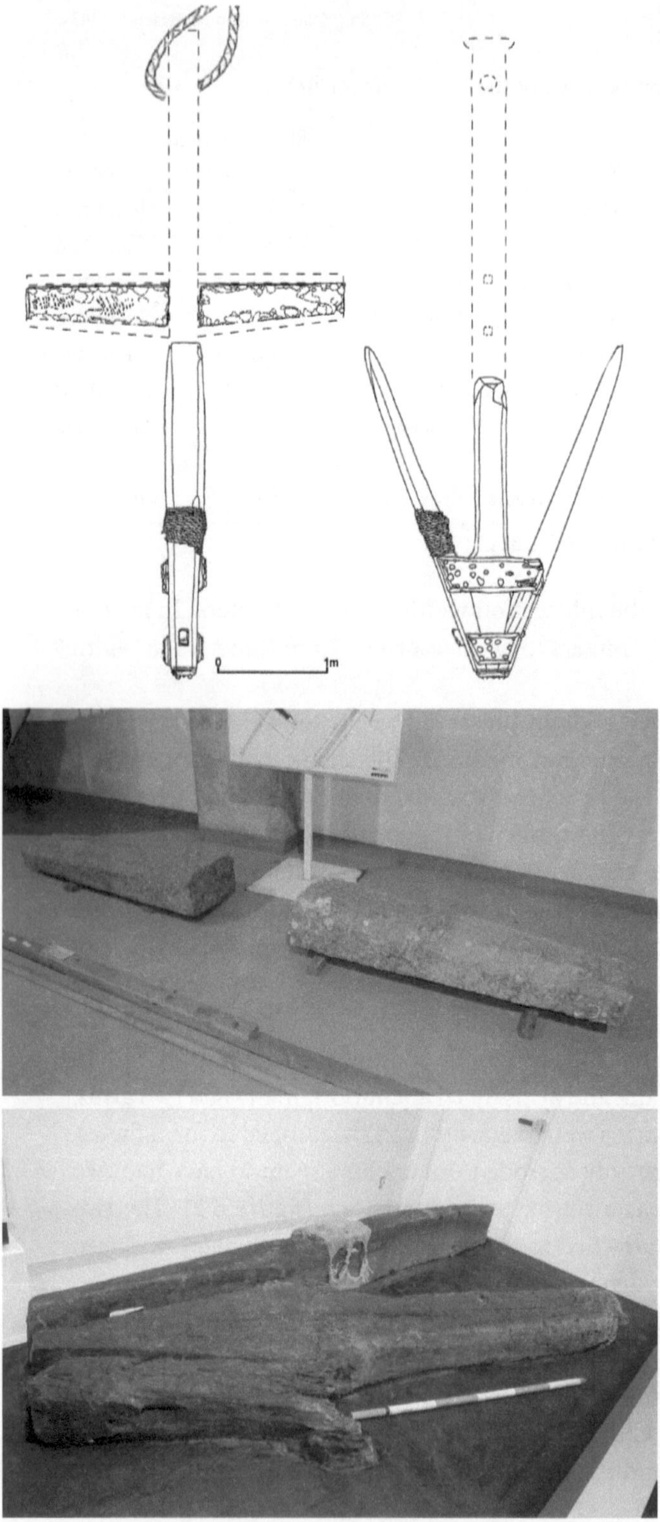

Figure 6.2. Drawing of anchor no. 3 from the Takashima underwater site (*top*) and photographs of a set of stone anchor stocks (*center*) and the wooden anchor (*bottom*) after conservation treatment with polyethylene glycol. (After Takashima-cho Board of Education 1996; photos by author)

Figure 6.3. Shank of a compound anchor stored in the Museum of Overseas Communication History. (Photo by author)

anchors using a single stone stock that penetrated the shank of the wooden anchor. Later, they started to use two-piece stone stocks. Both two-piece and single stocks appear to have been used concurrently throughout the thirteenth century. Yet the use of the single stone anchor stock can be confirmed during a much later period, according to an iconographical source depicting later Chinese ships (Oba 2003).

One of the characteristics of the compound anchor is the use of a stone stock positioned in the middle of the shank. Researchers have pointed out that traditional Chinese anchors have been distinguished by the location of the stock on the crown (Worcester 1966; Needham 1971; Worcester 1971). However, the compound anchors used during the Song and Ming dynasties show the installation of the stone stock in the midsection of the shank. Two large wooden anchor shanks in the Quanzhou Maritime Museum (Museum of Overseas Communication History) have two holes, possibly to enable the installation of stakes to hold a stone stock in the middle of the shank (figure 6.3). Another similar anchor was discovered in one of four shipwrecks excavated at Penglai Forts in Shandong Province, China. The ship is dated to the end of the Yuan dynasty or the early Ming dynasty (AD 1368–1644), according to the date of celadon recovered from the wreck and the construction date of the fort's sluice gate. The wooden anchor recovered from this shipwreck during the 1984 excavation season has two holes in the middle of the shank. Compared to these anchors, the anchors found at the Takashima underwater site provide the earliest-known archaeological evidence of the use of stone stocks in the East Asian region. The result of the most recent radiocarbon dating analysis of a specimen from the shank of anchor no. 3 from Takashima dated the anchor to circa AD 1269 (Takashima-cho Board of Education 2001: 51).

Scientific analysis of the stone stock of anchor no. 3, which is made of granite, has revealed that the chemical composition of the stone is similar to that of granite from Quanzhou regions in China (Suzuki et al. 2000). Moreover, wood used for the shank of anchor no. 2 originated in the region close to Quanzhou. These details offer minor clues as to where the ships equipped with the anchors originated.

Two hulls found more recently at the Takashima underwater site warrant mention. In 2012 a Tokai University team reported a buried hull identified in a remote sensing survey; details were presented at a UNESCO regional meeting in Koh Kong, Cambodia, on underwater cultural heritage. Since then, underwater excavation has been conducted at the remnants of the hull by Ryukyu University, with two ships having been detected by 2015. But archaeological reports regarding the newly discovered hulls were not available at the time of writing. Although the timbers previously discovered at Takashima were not in good enough condition to permit hull investigations to be performed on them, the data available in the published reports nevertheless have value as archaeological resources for further analysis of late thirteenth- and fourteenth-century East Asian vessels.

Iron Nails in East China Sea Ships

The use of iron fastenings is not solely a Chinese innovation; a few shipwrecks in the Mediterranean provide evidence of early adoption of iron fastenings during the Classical period (Steffy 1994; Kabanov and Pomey 2004; Duivenvoorde 2014; McCarthy 2005; Beltrame and Gaddi 2007). Recent archaeological discoveries of anchors associated with Southeast Asian indigenous ships indicate that ships from this region could have begun to use iron for anchors much earlier than was previously thought (see chapter 8). Iron fastenings have, however, long been used in Chinese

Table 6.2. Iron production methods developed in early China

Technology	Description
Sponge iron	The technology appears to have been developed around the ninth century BC and is the oldest iron production method. The technology is distinguished from the three that follow in that it was produced by the direct iron production method of low temperature reduction at 700°C. The metallography structure contains nonmetallic inclusions consisting of oxidized iron (FeO) and olivine (2FeO SiO_2) and shows large coarse ferrite.
Malleable cast iron	The technology appears to have been developed probably by the fifth century BC in China. The material was manufactured by annealing white cast iron for decarbonization at a temperature of 900–950°C. The metallography structure contains nonmetallic inclusions consisting of iron sulfides (FeS) or manganese sulfides (MnS) with grains of 5–10 μm and shows graphite grains with porosity from decarbonization, as well as ferrite and pearlite.
Iron casting decarbonized steel	The technology appears to have been developed probably by the third century BC in China. The material was manufactured by smelting pig iron and pouring it into square molds at a temperature of about 1,000°C in contact with air. After solidifying, it is annealed further and the decarbonization process occurs. The metallography structure contains nonmetallic inclusions consisting of iron sulfides (FeS) or manganese sulfides (MnS) with grains of 5–10 μm and shows ferrite and pearlite in some cases with porosity from the decarbonization process.
Paddling steel (wrought iron)	The technology appears to have been developed probably by the first century BC in China. The material was manufactured by smelting pig irons in the furnace, and it was paddled to be in contact with oxygen in the air for decarbonization and was then further hammered to exclude contaminations. The metallography structure contains nonmetallic inclusions consisting of silicate (Si, Al, Ca, Mg, K, Na) with grains of 20 μm and shows fine-grain ferrite.

Source: Osawa 2004.

shipbuilding. China had a long-term involvement in iron manufacturing, producing various types of iron, probably beginning with the production of cast iron. Different methods of iron production in early China are presented in table 6.2 (Osawa 2004). Iron is the most important metal used in East Asian shipbuilding technology. Iron nails characterize the East Asian shipbuilding tradition and distinguish it from Southeast Asian shipbuilding technology (Manguin 1993). The production of iron in China affected the development of iron manufacturing in Korea and Japan and showed progress during the tenth to twelfth centuries (Hartwell 1962, 1966; Wagner 1995), a period of innovation that allowed the Chinese to produce massive amounts of iron.

Iron Manufacturing and the Growth of Shipbuilding

During the Song dynasty (AD 960–1279), many larger furnaces began to be used in iron production facilities that were already using shaft furnaces, or blast furnaces (Wagner 2001). In the furnaces for smelting iron ore, coal was an essential factor, probably dating back to the period of the Northern Song dynasty (AD 960–1127). According to Robert Hartwell's historical study (1962), innovations in the brick and tile production industries permitted the replacement of firewood by mineral fuels in furnaces during the period. Over the following centuries, using coal to fire forges and furnaces became common. Along with these innovations, blast furnaces came to be used in iron manufacturing in the northern part of China (Hartwell 1962). Chinese socioeconomic history shows maturation of the technology throughout the Song dynasty in particular. During the period, various other industries grew, demanding a larger amount of iron. For example, during the twelfth century, copper coins were essential in the developing monetary system, but iron was technically still needed for copper production (Hartwell 1962). Shortly after the move of the Song capital south, 1,400 tons of iron were consumed annually in the hydrometallurgical process for removing copper from mine waters—a technique that involved boiling copper sulfate solution in contact with iron, the iron displacing the copper from its salt.

Northern China, around what is now Hebei Province, and areas toward the center of China, around Shandong Province and the northern part of Jiangsu Province, were focal areas for iron production. Archaeological evidence indicates that many furnaces were established in some of these areas (Wagner 2001). Significant iron production in northern China during the twelfth and thirteenth centuries has also been reported (Sasada 2009).

Whether iron manufacturing in the southern parts of China was as productive as in the north is arguable. There appear to have been some iron-forging facilities in the cities of Zhejiang Province, such as Mingzhou (Ningbo), and a few iron-production workshops in Fujian Province.

Despite the demand for iron for military purposes in Jiangsu and Zhejiang Provinces, facilities in these regions could not supply sufficient amounts. While the southern iron manufacturers may have been dominant suppliers, there appears to have been room for them to grow as consistent local suppliers in the south. Iron from the Fujian Province and farther south, in an area previously known as Guangnan, was traded in the markets of Jiangsu and Zhejiang Provinces (Shiba 1968: 300–301). The iron industry in Quanzhou was fully developed by the ninth century (Clark 1991: 28). Besides iron itself, iron products could also have been shipped from the south to these areas (Shiba 1968: 301). Yoshinobu Shiba pointed out that several districts in Quanzhou in Fujian Province were engaged in iron production, relying on manufacturing iron sand, and their iron products were sold in the provinces of Jiangsu and Zhejiang (Shiba 1968: 301). During the Yuan dynasty (AD 1271–1368), the iron trading network of Fujian and Guangnan expanded to the South China Sea. A few iron products including iron pots and needles from Quanzhou appeared on the list of trading commodities for Cambodia (Zhenla), evidenced in the *Zhenla Fengtuji* (Customs of Cambodia, 真臘風土記 [Shiba 1968: 302]).

Iron Nail Production and Shipbuilding

A historical text describes the involvement of the Song dynasty's iron industries in shipbuilding. Hartwell (1962: 161) refers to a Song dynasty record from the eighteenth-century Chinese historiography *Xu Zizhi Tongjian Changbian* (A sequel to annals of Chinese dynasties, 續資治通鑑長編), which says, "Shipbuilding at Shan-chou [*sic*; Cangzhou], in southern Hopei [*sic*; Hebei] along the banks of the Pei River, required supplies of iron and coal." This account does not clarify the purpose for which iron was to be used, but shipbuilding industries in Hebei coincided with iron industries in a relationship of consumer and supplier. Similarly, an account in *Song Huiyao Jikao* (Draft recompilation of the important documents of the Song, 宋會要輯稿) notes that "like the salt works, shipyards often had iron mines and smelters directly attached to them" (Hartwell 1962: 159). Additionally, "during the later years of the Northern Sung, these foundries supplied not only anchors and nails, but also iron prows [*sic*] and other armor" (Hartwell 1962: 159). An example of an iron-forging workshops placed within a shipyard is evidenced in the Ming dynasty's shipyard in Nanjing (Church 2010).

How intensively iron industries were linked with shipbuilding industries is important for understanding the availability of nails. In terms of the larger perspective, we need to consider the impact of the Song dynasty's development of iron industries on the changing shipbuilding tradition in the southern regions. Some sources imply that south of Quanzhou, iron industries may have had less influence. According to a description appearing in the Song dynasty historical text *Lingwai Daida* (or *Ling Wai Tai Ta*, Answers to the question about the areas beyond our territory, 嶺外代答), iron nails and tung oil were not easily available in Guangdong, where lashed ships predominated:

> "They always used hollow planks leashed with ratten," and dried madder stalks were used to calk the seams. (*Lingwai Daida*, vol. 6: *Teng chuan* [translated in Merwin 1977])
> 造船皆空板穿藤約束而成．於藤縫中，以海上所生茜草. (嶺外代答，卷6藤舟)

Chinese researchers have proposed that the lashed or sewn ship built without iron nails in Guangdong in southern China suggests that technologies in the south might have lagged behind those of northern and central China (Merwin 1977: 75–76). However, the text of *Lingwai Daida* continually mentions the excellent seaworthiness of the lashed ships with their tight caulking system. These ships appear to have been quite large, and people seem to have preferred them in the South China Sea. A contrasting tradition of constructing large ships without iron materials seems to have prevailed in parallel with the adoption of iron nails in many regions during the Song dynasty.

Iron Nails in East Asian Ship Remains

Most of the iron nails in the ship remains from the Quanzhou ship, the Shinan shipwreck, and the Takashima underwater site have not remained intact. X-ray fluorescence (XRF) analysis of timber from the Shinan shipwreck shows the oxidization of an iron nail, and its condition of degradation appears to correspond with a general understanding of corrosion on iron nails embedded in ship timbers (Huisman 2009; Sandström et al. 2002, 2005). Such degradation has meant a loss of the information necessary to obtain metallographic perspectives with a focus on the iron nail manufacturing processes. The following examination of remnants of iron nails from the two East Asian traders and Takashima underwater site is considered to be a pilot study for better understanding iron fastenings in the East China Sea shipbuilding tradition.

The Quanzhou Ship

During my investigation in 2009, the museum staff confirmed that the original iron nails did not remain in the hull. The original nails removed from the hull were displayed in the exhibition room, and they have since degraded. Few studies are useful for acquiring information about the original configuration of the nails (Xu 1985; Li 2004).

According to Guoqing Li (2004), when modern metal nails were used to replace the degraded nails, only two different types of iron nails were used. Their size is consistent with the originals in length: the smaller type measures about 150 millimeters, and the larger type varies from 200 millimeters to 220 millimeters. For assessment of the original nails, black and white photographs of nails removed from the hull proved helpful (figure 6.4): two less-degraded nails show a tapered configuration and have flat nailheads. The photographs do not show a scale, so determining the dimensions of the nails is difficult.

I thus could not undertake a metallurgical analysis to determine the quality of the iron remains in the Quanzhou ship. My examination was necessarily limited to details observed in the hull in its current condition. As mentioned previously (see chapter 4), the uppermost plank remains expose the seam of the hull planks, where corroded nails and nail holes are visible. Most of the nail holes are single square holes, but there are some double holes aligned together (figure 6.4). The reason for the double holes has not been clarified. From the limited information about the nails of the Quanzhou ship, we can conclude of their general configuration that the nail shaft has a square cross section and the taper is likely to occur only near the end.

The Shinan Shipwreck

No detailed study of the iron nails in the Shinan shipwreck was conducted previously. To identify magnetic remains, I attempted an examination during the investigation in 2009 in conjunction with staff of the National Research Institute of Maritime Cultural Heritage. Magnetic remains were

Figure 6.4. Degraded iron nails from the Quanzhou ship and the exposed seams of the uppermost planks showing nail holes (marked with white arrows). (Black and white nail photo courtesy of Museum of Overseas Communication History; seam photo by author)

barely identified in part of the hull of the shipwreck. This sign of possible magnetic remains in three ship timbers—a portion of a hull plank (SW-82-182; figure 6.5), a sheathing (sacrificial) plank, and a butt plate (figure 6.6)—warranted further investigation to determine whether this signaled the presence of iron remains and, if so, whether the iron was sufficient to permit metallurgical assessment.

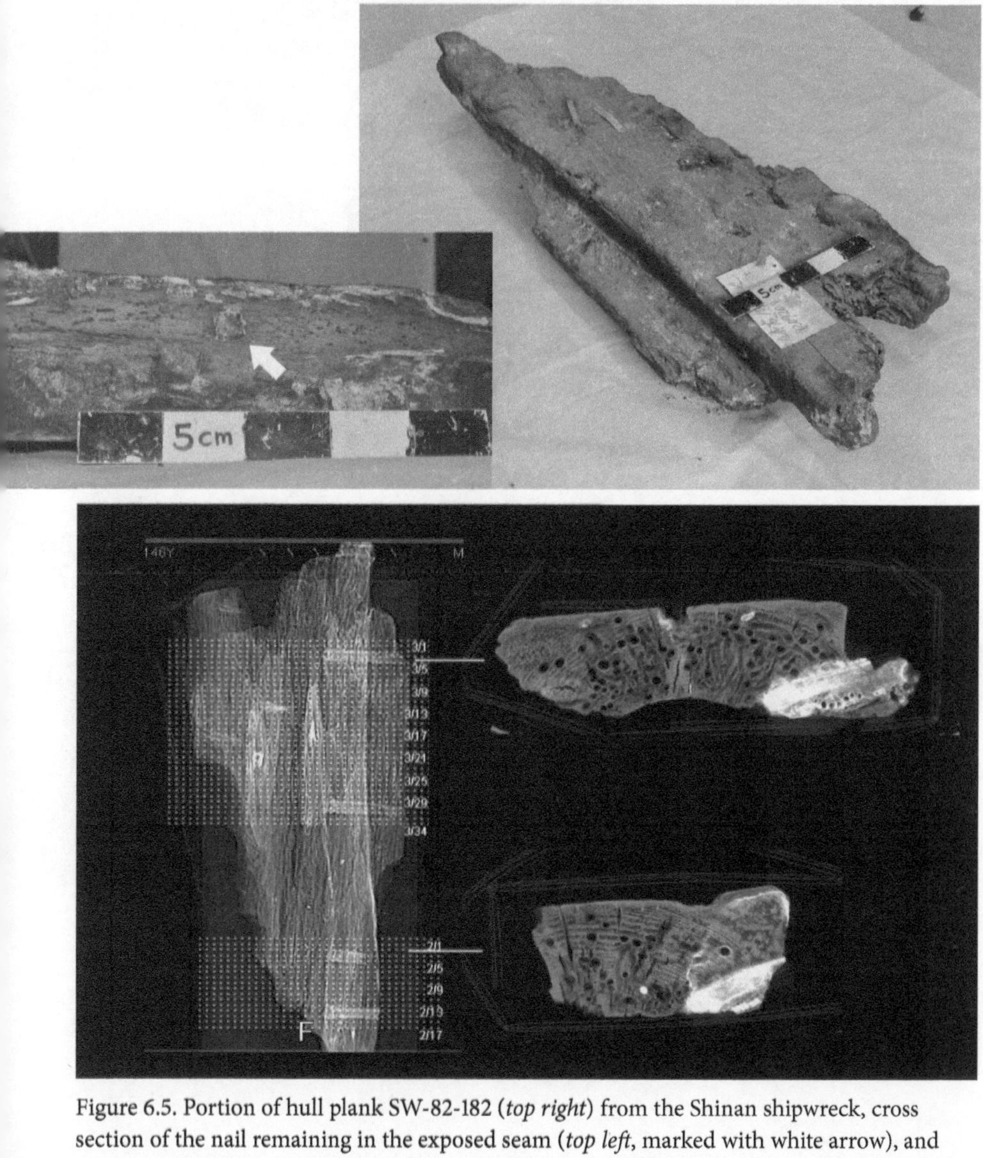

Figure 6.5. Portion of hull plank SW-82-182 (*top right*) from the Shinan shipwreck, cross section of the nail remaining in the exposed seam (*top left*, marked with white arrow), and CT scan image showing remains of the nail (*bottom*). (Scan courtesy of National Maritime Museum of Korea; photos by author)

Figure 6.6. Sheathing plank from the Shinan shipwreck (*top*) and CT scan image (*middle*) showing evidence of the use of five nails and butt plate (*bottom*) showing the location of nails. (Courtesy of the National Research Institute of Maritime Cultural Heritage)

X-ray and CT scan analyses are nondestructive approaches for revealing nail positions and driving patterns not visible from the outside. In this study, such analyses were conducted to determine the quality of iron remains. All the samples used were separated timbers that had not been restored. The analysis indicates that the quality of the corroded nails varies. Some X-ray and CT scan images present fully corroded iron remains, evident as a black hollow image with only a thin magnetite layer to indicate the original surface (figure 6.5). Some nail remains show white blurry images, which suggest possible mineral remains inside the nail. It is difficult to determine the detailed corrosion status from the white blurry images just by looking at the X-ray and CT scan images.

A specimen of timber cut from the hull plank (SW-82-182) was provided to Kyushu Techno Research Inc. for metallurgical study. The specimen was further cut to expose sections of the nail, and microscope images of the sections were produced. The microscopically observed structure shows that iron has been fully oxidized and the nail no longer contains metal structures. In its corroded state, the nail is represented only by hydrated iron oxide. The substantial oxidization caused the loss of the metal structure, which restricted the ability to obtain metallurgical information. However, interpretation of how the original nails had been manufactured was attempted based on the minimal information that can be seen in the microscope images (Ian Macleod, pers. comm., 2010). The images indicate casting procedures in manufacturing, as well as the structure of wrought iron, which appears as a circular distribution of inclusions. This indicates original working patterns of iron manufacture using a method involving folding and reforging: the metal was annealed (a process of heating and cooling back to air temperature to make metal more workable [see table 6.2]) and then forged, perhaps by hammering.

The Takashima Underwater Site

Metallurgical analysis was also attempted on corroded remains of two iron nails mentioned earlier from bulkhead planks discovered at the Takashima underwater site. The two nails have fully degraded; thin magnetite layers have formed, and the inside of the nail holes is hollow. There is thus no metallurgical information that would permit a determination of the manufacturing processes used for the original nails. Despite the limited information, based on his previous research Masami Osawa (2005a) inferred that the quality of the iron used for the nails was similar to what is known as iron casting decarbonized steel (see table 6.2).

According to Osawa (2005a, 2005b), the manufacturing process of the Takashima nails can be inferred through iron remains found at the Avraga site, which is regarded as Genghis Khan's mausoleum, dating to the thirteenth century. Thin rectangular ingots of cast iron were initially produced by using molds measuring 10–15 millimeters in cross section. The cast iron ingots were too brittle to be used for ship nails. Consequently, an annealing process must have been conducted—by the application of heat, probably for a few days, and by exposure to the air—whereby decarbonization of the iron occurred. As a result, the iron became soft and pliable. It could then be further forged, producing nails. The quality of iron produced through these processes is defined as an iron casting decarbonized steel. During the process, the composition of the iron changed, yielding an extra-soft steel (less than 0.001 percent carbon). Study of the iron products found at the Takashima underwater site has provided insights that help us determine the quality and manufacturing processes of the nails from the Quanzhou ship and the Shinan shipwreck.

Iron was a primary material used for fastenings in the construction of seagoing ships sailing in the East China Sea. This owed a great deal to the development of iron production in China, where industries were remarkably mature by the time of the Song dynasty. Apart from the need for clarification of the manufacturing process, the entire system of obtaining iron nails as a completed product is still unknown. The relationship between iron industries and shipbuilding industries clearly requires further study.

Timber Analysis

Descriptions of timber used for ship construction appear in some detail in a seventeenth-century Chinese historical text:

> On the original specification, the grain-ship is built 52 ft. long with planks 2 in. thick. The choicest timber for it is large baulks of nanmu, but the chestnut (li) is also used as second best. (*Tian Gong Kai Wu*, vol. 9 [translated in Needham 1971: 411])
> 糧船初制．底長五丈二尺．其板厚二寸．採巨木楠爲上．栗之次．(天工開物 第九)

> The timber for the mast is usually fir (shan-mu), which must be straight and sound. If the natural size of the spar is not long enough for the mast, two pieces can be coupled together by means of a series of iron bands placed around the joint a few inches apart. . . . Hull timbers and bulkheads are made of nan-mu, zhu-mu, camphor wood (zhang-mu), elm (yu-mu), or sophora wood (huai-mu). [Camphor wood, if taken from a tree felled in spring or summer, is liable to be attacked by boring insects or worms.] Deck planks can be made of any wood. The rudder-post is made of elm, or else of ye-mu or of zhu-mu. The tiller should be of chou-mu or of ye-mu. The oars should be of fir (shan-mu) or juniper (kuei-mu) or catalpa wood (qiu-mu). (*Tian Gong Kai Wu*, vol. 9 [translated in Needham 1971: 414; bracketed interpolation from Needham])
> 凡木色.　桅用植杉木.　長不足則接.　其表鐵箍逐寸包圍. . . . 梁興枋檣. 用楠木，櫧木，樟木，榆木，槐木. [樟木春夏伐採. 久則粉蛀.] 棧板不拘何木. 舵棹用榆木，榔木，櫧木，關門棒用稠木. 櫓用杉木，檜木，楸木. (天工開物 第九)

These descriptions indicate that while the wood preferably used for masts was fir (*Cunninghamia lanceolata*), relatively harder woods were used for the main components of the hull, such as bay or *nan mu* (*Machilus* sp.), oak or *zhu mu* (*Quercus* sp.), camphor or *zhang mu* (*Cinnamomum* sp.), elm or *yu mu* (*Ulmus* sp.), and sophora or *huai mu* (*Styphnolobium* sp.). While there are no specific records for the timbers used in the superstructure of the ships, those used for components of a ship's steering system would have been carefully selected. Joseph Needham (1971: 645) introduces an early account of the selection of timbers in a third-century historical text, *San Fu Huang Guo*, referring to an emperor's order for the hull of a small boat to be constructed using catalpa wood, with magnolia wood for *yulohs*. Shiba (1968: 223) has pointed out that pine, fir, and camphor were the predominant trees used for ship construction during the Song dynasty. Archaeological evidence supplements the historical information, with timber species identification data for the Quanzhou and Shinan ships and for the ship remains from the Takashima underwater site indicating the regional locations of shipyards where they were built. The discussion demonstrates the usefulness of species identification in the study of hull components, especially in comparative perspective.

Timber Species Identification

J. Richard Steffy (1994: 256–59) in his nautical archaeological study provides a discussion of major trees used in European shipbuilding. For the study of wood species in East Asia, databases specific to the region are pertinent, including the online database "Flora of China," providing botanical information about China, and online data sets provided by the Forestry and Forest Products Research Institute in Japan are also a useful source for identifying the kinds of timber available in Japan and Southeast Asia. Information from these sources proved valuable in analyzing the wood in East Asian ship remains. Primary data for timber species identifications and the anatomical properties of wood in the hull structural remains are available for each site in the reports published by the holding organizations and institutions (Museum of Overseas Communication History 1987; National Maritime Museum of Korea 2004; Takashima-cho Board of Education 2008). These reports allow us to understand the norms of shipbuilding in the selection of specific woods for each hull component, related to utilitarian factors of the species (table 6.3).

The Quanzhou Ship

Wood species identifications were conducted on samples from selected parts of the hull of the Quanzhou ship in 1974, 1975, and 1978 (Museum of Overseas Communication History 1987). Sixteen specimens were assessed from nine hull components: the main keel, the forward keel, the aft keel, frames, a windlass, bottom ceiling planks, garboards, hull planks, and bulkheads. The specimens were compared to modern wood specimens provided by the Forestry Agency of Jinjiang district and Jinjiang Timber Company in Yongchun, Fujian Province, for the identifications. All the samples were taxonomically identified to the species level, and each sample was classified into one of three species: *Pinus massoniana* Lamb., known as Chinese pine; *Cunninghamia lanceolata* (Lamb.) Hook., known as Chinese fir; and *Cinnamomum camphora* (L.) J. Presl, known as the camphor tree.

Wood species identifications on three timber samples from the Quanzhou ship were conducted by the Forestry and Forest Products Research Institute in 2009 (see chapter 4 for radiocarbon dating of two of these samples). The most recent species identification analyses confirmed the use of *Cunninghamia lanceolata* and *Cinnamomum camphora* (Noshiro

Table 6.3. Types of woods used for the timbers from the Quanzhou ship, Shinan shipwreck, and Takashima underwater site

Hull component	Common name	Botanical name	Chinese description
Quanzhou ship			
Main keel	Chinese pine	*Pinus massoniana* Lamb.	Ma wei song (马尾松)
Forward keel	Camphor	*Cinnamomum camphora* (L.) J. Presl	Xiang sha (樟树), xiang zhang (香樟)
Aft keel	Camphor	*Cinnamomum camphora* (L.) J. Presl	Xiang sha (樟树), xiang zhang (香樟)
Frame	Camphor	*Cinnamomum camphora* (L.) J. Presl	Xiang sha (樟树), xiang zhang (香樟)
Windlass	Camphor	*Cinnamomum camphora* (L.) J. Presl	Xiang sha (樟树), xiang zhang (香樟)
Ceiling plank	Camphor	*Cinnamomum camphora* (L.) J. Presl	Xiang sha (樟树), xiang zhang (香樟)
Garboards	Chinese fir	*Cunninghamia lanceolata* (Lamb.) Hook.	Shan mu (杉木), sha mu (沙木)
Planks	Chinese fir	*Cunninghamia lanceolata* (Lamb.) Hook.	Shan mu (杉木), sha mu (沙木)
Bulkheads	Chinese fir	*Cunninghamia lanceolata* (Lamb.) Hook.	Shan mu (杉木), sha mu (沙木)
Shinan shipwreck			
Main keel	Chinese pine	*Pinus massoniana* Lamb.	Ma wei song (马尾松)
Forward keel	Chinese pine	*Pinus massoniana* Lamb.	Ma wei song (马尾松)
Aft keel	Chinese pine	*Pinus massoniana* Lamb.	Ma wei song (马尾松)
Garboards	Chinese pine	*Pinus massoniana* Lamb.	Ma wei song (马尾松)
Planks	Chinese pine	*Pinus massoniana* Lamb.	Ma wei song (马尾松)
Bulkheads	Chinese pine	*Pinus massoniana* Lamb.	Ma wei song (马尾松)
Sheathing (sacrificial) planks	Chinese fir	*Cunninghamia Lanceolata* (Lamb.) Hook.	Shan mu (杉木), sha mu (沙木)
Part of water tank	Chinese fir	*Cunninghamia Lanceolata* (Lamb.) Hook.	Shan mu (杉木), sha mu (沙木)
Mast step	Camphor	*Cinnamomum* sp. (*Cinnamomum camphora* [L.] J. Presl?)	Xiang mu (樟木)
Water tank bracket	Winter-hazel	*Distylium* sp.	Wen mu shu (蚊母树)

(continued)

Table 6.3—*Continued*

Hull component	Common name	Botanical name	Chinese description
Unknown timber associated with keel from join of main keel and forward keel	Oak	*Cyclobalanopsis* sp.	Qing gang (青冈)
Takashima underwater site			
Anchor no. 2: shank (no. 15)	Holly olive	*Osmanthus heterophyllus* (G. Don) P.S. Green	Yi ye dong shu (异叶冬树)
Anchor no. 2: arm (i—no. 21)	Oak	*Cyclobalanopsis* sp.	Qing gang (青冈)
Anchor no. 2: arm (ii—no. 51)	N/A	Broadleaf tree (Unidentified)	N/A
Anchor no. 2: wooden stakes to hold stock (nos. 10 & 13)	Oak	*Cyclobalanopsis* sp.	Qing gang (青冈)
Anchor no. 2: loose tenon (no. 52)	N/A	Broadleaf tree (Unidentified)	
Anchor no. 3: shank (no. 24)	Oak	*Cyclobalanopsis* sp.	Qing gang (青冈)
Anchor no. 3: arm (i—no. 54)	Oak	*Cyclobalanopsis* sp.	Qing gang (青冈)
Anchor no. 3: arm (ii—no. 28)	Oak	*Cyclobalanopsis* sp.	Qing gang (青冈)
Anchor no. 3: wooden stake to hold stock (no. 27)	Oak	*Cyclobalanopsis* sp.	Qing gang (青冈)
Anchor no. 3: loose tenon on anchor crown	Oak	*Cyclobalanopsis* sp.	Qing gang (青冈)
Anchor no. 3: trapezoidal wooden plates on anchor crown	Camphor	*Cinnamomum camphora* (L.) J. Presl	Xiang sha (樟树), xiang zhang (香樟)
Anchor no. 4: shank (no. 44)	Oak	*Cyclobalanopsis* sp.	Qing gang (青冈)
Anchor no. 4: arm (i—no. 47)	N/A	Broadleaf tree (unidentified)	N/A

Hull component	Common name	Botanical name	Chinese description
Anchor no. 4: arm (ii—no. 56)	Fig	*Ficus* sp.	Rong mu (榕木)
Anchor no. 4: wooden stake to hold stock (nos. 41 & 42)	N/A	Broadleaf tree (unidentified)	N/A
Anchor no. 4: loose tenon	N/A	Broadleaf tree (unidentified)	N/A
No. 8 (small ship timber)	Camphor	*Cinnamomum camphora* (L.) J. Presl	Xiang sha (樟树), xiang zhang (香樟)
No. 15 (large ship timber)	Camphor	*Cinnamomum camphora* (L.) J. Presl	Xiang sha (樟树), xiang zhang (香樟)
No. 18 (large ship timber)	Camphor	*Cinnamomum camphora* (L.) J. Presl	Xiang sha (樟树), xiang zhang (香樟)
No. 19 (bulkhead?)	Camphor	*Cinnamomum camphora* (L.) J. Presl	Xiang sha (樟树), xiang zhang (香樟)
No. 193 (mast step?)	Camphor	*Cinnamomum camphora* (L.) J. Presl	Xiang sha (樟树), xiang zhang (香樟)
KZK02 no. 221 (thin ship timber)	N/A	N/A	N/A
KZK02 no. 601 (hull plank)	Camphor	*Cinnamomum camphora* (L.) J. Presl	N/A
KZK02 no. 909 (hull plank)	Camphor	*Cinnamomum camphora* (L.) J. Presl	Xiang sha (樟树), xiang zhang (香樟)
KZK02 no. 949 (hull plank)	Camphor	*Cinnamomum camphora* (L.) J. Presl	Xiang sha (樟树), xiang zhang (香樟)
KZK02 no. 1142 (bulkhead or plank?)	Camphor	*Cinnamomum camphora* (L.) J. Presl	Xiang sha (樟树), xiang zhang (香樟)
KZK02 no. 1236 (bulkhead near nos. 1439 and 1440)	N/A	N/A	N/A

(*continued*)

Table 6.3—*Continued*

Hull component	Common name	Botanical name	Chinese description
KZK02 no. 1439 (bulkhead plank)	Camphor	*Cinnamomum camphora* (L.) J. Presl	Xiang sha (樟树), xiang zhang (香樟)
KZK02 no. 1440 (bulkhead plank)	Camphor	*Cinnamomum camphora* (L.) J. Presl	Xiang sha (樟树), xiang zhang (香樟)
KZK02 no. 1476 (frame?)	Castanopsis	*Castanopsis* sp.	Zhu li (锥栗)
KZK02 no. 1607 (thin ship timber)	Pine	*Pinus* sp.	Song mu (松木)
KZK02 no. 1863 (plank)	Castanopsis	*Castanopsis* sp.	Zhu li (锥栗)
TKS13 no. 8 (athwart beam or bulkhead)	Camphor	*Cinnamomum camphora* (L.) J. Presl	Xiang sha (樟树), xiang zhang (香樟)
TKS13 no. 16 (plank)	N/A	Taxodiaceae	N/A
TKS13 no. 17(plank, joining to no. 16)	Pine	*Pinus* sp., subgenus Diploxylon	N/A
TKS14 no. 26 (bulkhead)	Camphor	*Cinnamomum camphora* (L.) J. Presl	Xiang sha (樟树), xiang zhang (香樟)

Source: Museum of Overseas Communication History 1987; National Maritime Museum of Korea 2004; Takashima-cho Board of Education 2008.

and Abe 2010)—the same species that were identified in previous research (Museum of Overseas Communication History 1987). The configuration of the samples used for analysis, however, does not provide evidence indicating which part of the hull these timbers came from. The sections of the sample identified as *Cunninghamia lanceolata* show substantial damage by marine borers, suggesting that the original timber had been exposed to water and thus perhaps was a hull plank. The sample identified as *Cinnamomum camphora* might have been from a plank of the bulkhead or fairing laths, if we consider its overall dimensions and look at its cut edge. Although the information is insufficient to permit a definite determination of the origin of the timbers, the results of the recent species identifications do not contradict the wood species assemblage previously identified for the hull, in which two types of woods were used as primary construction materials.

The Shinan Shipwreck

Korean researchers conducted wood species identifications on hull components of the Shinan shipwreck, and the results are available in reports (National Maritime Museum of Korea 2004, 2006b). Identifications were conducted on the main hull components: the keel, garboards, planks, and bulkheads. All the specimens from these main structural components were identified as *Pinus massoniana*, which was apparently the primary construction material for the Shinan ship.

The growth rings of the timbers measure 4–8 millimeters on average, indicating that the timbers had come from trees in environments favorable for significant growth. Some clear rings are distinctive, indicating substantial spring seasonal growth. *Cunninghamia lanceolata* was identified for a few parts of the hull, including the sheathing planks and part of the water tank. Degradation of the timbers made from this tree varies; some have deteriorated, while others still show well-preserved cell structure. The growth rings of the timbers measure 2 millimeters on average, and ring width is consistent. The mast cheeks were identified as *Cinnamomum* sp. The report notes that this genus has various species, and the species could not be identified. However, it is likely to be *Cinnamomum camphora*, given the many other examples demonstrating that this wood was the camphor species most commonly used for shipbuilding in the region (National Maritime Museum of Korea 2004). Wooden brackets from the water tank were identified as *Distylium* sp. The timber shows a soft structure, and the ring width measures only 0.4–0.6 millimeters. *Distylium* is a genus in the hazel family (Hamamelidaceae), in which about eighteen species are native to China. *Distylium racemosum* is likely to be the species used for the brackets, but that has not been confirmed definitively. *Cyclobalanopsis* sp. was used for a small rectangular timber from the join of the forward and main keel. Although the purpose of this timber has not been clarified, *Cyclobalanopsis* sp. is commonly used for hull structures and rigs in the region (National Maritime Museum of Korea 2004). Thus, most of the species found in the hull of the Shinan shipwreck were coniferous trees that originated in the southern parts of China (and Taiwan). Timber from well-grown *Pinus massoniana* trees composed the hull. Trees that had reached around 150–170 years old were used for the hull (particularly the keel), and trees aged around 60–100 years were used for most other main hull components (National Maritime Museum of Korea 2004: 23).

The Takashima Underwater Site

Data on wood species recovered from the site are available in several reports (Takashima-cho Board of Education 1996, 2001, 2003, 2008). The early wood species identifications on the large wooden anchors are important (Mitsutani 1996). Of the nine anchors found, the results of species identifications on the three that have the fullest set of components (anchors nos. 2, 3, and 4) are summarized in table 6.3. The structure of these anchors shows consistency, yet different taxa of woods were identified: *Osmanthus heterophyllus* (G. Don) P. S. Green (anchor no. 2) and *Quercus* sp. and *Cyclobalanopsis* sp. (anchors nos. 3 and 4). The *Osmanthus heterophyllus* used for the shank of anchor no. 2 is notable in that it is endemic to Taiwan. This fact should be considered in discussing the origin of Kublai Khan's fleet; some ships in the fleet might have been constructed in or near Taiwan. As mentioned early in this chapter, chemical analysis conducted on the stone anchor stocks suggests that the stone probably originated in the coastal areas of China facing the Taiwan Strait. The endemic species range is not direct evidence of the place of construction, however, because wood may have been exported by timber industries in Taiwan.

The species identifications demonstrate that anchor components may have been made from different woods. The shank and arm of anchors nos. 2 and 4, for example, were made of different woods. While the same species of wood was used for the shank and arm of anchor no. 3, other components of this anchor were made from *Cinnamomum camphora*.

Timber species identifications were conducted on a total of 606 specimens from many wooden remains recovered during the 2002 rescue excavations and in test excavations in the following years (Takashima-cho Board of Education 2008). Of these, I focus in table 6.4 on the species identifications for 237 specimens classified as ship timbers. These identifications demonstrate a wide range consisting of twenty-six wood taxa, of which only two—*Cinnamomum camphora* and *Cunninghamia lanceolata*—could be identified to the species level. In total, 104 ship timbers are identified as one of these two species: 63 are *Cinnamomum camphora* and 41 are *Cunninghamia lanceolata*. The third-greatest number is 34 samples of *Pinus* sp. (subgenus *Diploxylon*). These three woods serve as major

Table 6.4. Assemblage of wood species from the Takashima underwater site

Species	Number	Percentage of total
Cinnamomum camphora	63	27
Cunninghamia lanceolata	41	17
Pinus sp., subgenus *Diploxylon*	34	14
Ulmus sp.	17	7
Taxodiaceae	17	7
Lauraceae	14	6
Pinus sp., subgenus *Haploxylon*	14	6
Cryptomeria sp.	6	3
Castanopsis sp.	6	3
Other	25	11
Total	237	

Source: After Takashima-cho Board of Education 2008.

trees that constitute 58 percent of the assemblage of wood remains classified as ship timbers.

The assemblage of other wood remains (42 percent of the specimens) consists of twenty-three taxa. Among the miscellaneous timbers discovered at the Takashima underwater site, there appears to be a correlation between timber size and wood species; the timber remains originally having large dimensions, indicative of their use for ship construction (including the bulkhead planks), tend to be identified as *Cinnamomum camphora*.

Timber for Shipbuilding in the East China Sea Region

George R. G. Worcester (1941) introduced the names of woods used to build native river craft in China. The ships plying the upper Yangtze River were built using *song mu* (pine), *qing gang* (oak), *nan mu* (*Machilus* sp.), *hong chun* (camellia), *feng xiang* (maple), or *bai mu* (cypress). Flora is varied in such a large continent, and the types of woods used in shipbuilding could have changed through time. Nevertheless, the wood remains of excavated ships dating to the late thirteenth and early fourteenth century demonstrate selection of particular types of wood: camphor tree, *Cinnamomum camphora*; Chinese pine, *Pinus massoniana*; and Chinese fir, *Cunninghamia lanceolata*. These three species were the woods most commonly used during the growth period of East China Sea shipbuilding. Some botanical information about these species warrants inclusion.

The camphor tree, *Cinnamomum camphora* (L.) J. Presl, grows to thirty meters tall with a trunk diameter of up to three meters. It is widely cultivated south of the Yangtze River (Chang Jiang) and also grows in Japan and part of Korea (Jeju Island). This tree yields oleoresin, already known as *chang nao* during the Song dynasty (*chun nao*, 樟脑 [Ptak 2000]), though how the utility of such a substance was regarded in early Chinese shipbuilding is unclear. In Japan, the camphor tree was used for shipbuilding as commonly as *Cryptomeria japonica* (Japanese cedar); in Korea the restricted use of camphor is evidenced by the species identifications from ships of Korean origin.

The Chinese pine, *Pinus massoniana* Lamb., grows to a large size appropriate for ship construction: forty-five meters tall with a trunk diameter of up to one and a half meters. Important in timber industries, it is widely cultivated in modern China. Its distribution is in the south, including Anhui, Fujian, Guangdong, Guangxi, Guizhou, Hainan, western Henan, Hubei, Hunan, southern Jiangsu, Jiangxi, southeastern Shaanxi, Sichuan, eastern Yunnan, and Zhejiang Provinces. This is a southern native species of mainland China and Taiwan. In Japan and Korea it is not as common as in China.

The Chinese fir, *Cunninghamia lanceolata* (Lamb.) Hook., grows to fifty meters tall with a trunk diameter of up to three meters. It is an indigenous Chinese species, and its distribution is similar to that of the camphor tree: Anhui, Fujian, Gansu, Guangdong, Guangxi, Guizhou, Hainan, Henan, Hubei, Hunan, Jiangsu, Jiangxi, Shaanxi, Sichuan, Yunan, and Zhejiang Provinces. The "Flora of China" database states that this wood is resistant to rot and termites and is easily worked. These characteristics explain its use for sheathing planks in the Shinan shipwreck; sheathing planks are thin and shaped to various sizes and function to counter biotic reactions degrading wood tissues, including the effects of shipworm. This wood shows remarkable dominance among the species identified in the Ming dynasty's Treasure Shipyard in Nanjing, which was a focal shipyard providing ships to Zheng He's expedition fleet (see chapter 3).

It is also worth looking at the hardness of these timbers. Use of hardwoods is likely to correlate with the sturdiness of the ship structure. Hardwoods here are defined as broad-leaved trees, or flowering trees, in comparison to coniferous trees, which are defined as softwoods. The term *hardwood* means dense or heavy wood. In general, an increase in a wood's weight is in proportion to the increase in hardness. Attributes for

Table 6.5. Density and hardness of the woods

Species		Brinell hardness		
	Air-dried wood density (g/cm^3)	Cross section	Radial section	Tangential section
Cinnamomum camphora	0.53–0.58	4000–4200	3100–3500	3200–3670
Pinus massoniana	0.44–0.59	2520–4220	2080–3220	2540–3600
Cunninghamia lanceolata	0.32–0.39	2140–3040	1280–1850	1540–2060

determining the weight of woods include the density of the cells and the amount of lignin in the cell walls. Wood, however, has many air spaces in the cell lumen (pores), between cells, and between cell walls, and the percentage of these spaces depends on the type of wood. Without these spaces, pure wood cell wall material shows a density of about 1.50 grams per cubic centimeter (g/cm^3), yet such a wood does not exist, because plants need spaces to carry or store water, nutrients, and gas for biological processes. Thus, 1.50 grams per cubic centimeter is a weight defined as the specific gravity of wood, such that wood with a weight close to 1.50 grams per cubic centimeter is considered a heavier wood.

The relative weight of different woods based on specific gravity can be measured by calculating the amount of gas filling up the spaces in completely dried wood. The resulting value representing the specific gravity of the dried wood, which is equalized to a volume of pure water, is known as the relative density of dry wood, or the air-dried wood relative density. The relative density of dried wood of the three focal species (table 6.5) is drawn from a Chinese source (Cheng 1985: 1258–94). For *Cinnamomum camphora*, the air-dried wood density values presented in the Japanese wood database are slightly different: 0.41–(0.52)–0.69.

Another criterion for assessing hardness of wood is its Brinell hardness. The Brinell scale is used as one of the standards for defining hardness of materials in industry by measuring the penetration of a weight loaded on the material. The Brinell hardness of key woods is also presented in table 6.5.

Because the specific gravity values correspond to one gram of pure water, wood with a relative density value greater than 1.00 theoretically does not float in pure water, depending on the condition of the wood. In terms of specific gravity, oak, for example, is regarded as a heavy wood, and on average it shows a relative density of 0.90. In contrast, the relative density values for the camphor tree and Chinese pine would classify them as medium-heavy woods that also provide appropriate hardness. The results of species identifications for the Quanzhou ship and the Shinan shipwreck suggest that the hardness of those two species is sufficient for wood from those trees to be serviceable in important hull components. Chinese fir is a lighter tree consisting of softer wood tissue, and it is a convenient tree for mass production of ship timbers. Wood from Chinese fir appears to have been used for hull planking.

Castanopsis sp., identified for a few timbers from the Takashima underwater site and regarded as ship construction material, shows a relative density similar to that of the camphor tree and Chinese pine (about 0.60 on average). Some of the other wood remains from the sites are identified as heavier woods, such as *Cyclobalanopsis* sp., generally regarded as a lower-level taxon of oak (a subgenus of *Quercus*). This wood was used for timbers associated with the keel of the Shinan shipwreck and the shanks of two Takashima anchors (anchors nos. 3 and 4). Anchor no. 3, the largest anchor among those found, consists mainly of this relatively heavy wood. Woods used for constructing hulls vary in air-dried wood density, with values ranging from 0.30 to 0.60, depending on the component. For the anchors to sink properly, oak or heavier woods would presumably have been preferred.

Wood Species Identifications for Other East Asian Ships

The types of wood used in East Asian ship construction have been of interest for a long time, yet the ship remains discovered circa the 1970s–1980s provided only limited information on that topic. For the early-discovered Ningbo shipwreck, for example, the types of wood used for the hull are merely noted: Chinese fir, Chinese pine, and camphor tree (Green 1997a). More information has recently become available, however, to facilitate comparative study of wood species. Wood species identifications have been conducted on components of the Penglai ships of the Yuan to Ming dynasties (see chapter 7 for more about these ships). Penglai ship no. 1 was discovered in 1984, and only a short summary of species identifications is available (Liu et al. 2006). The Research Institute of Wood Industry of the China Academy of Forestry assessed the wood species in the remains of the other three ships in 2005, and the details have been reported (Liu et al. 2006), as summarized in table 6.6.

During the latest identifications, forty-nine specimens from hull components of Penglai ships nos. 2, 3, and 4 were recovered and analyzed. Together with woods identified from Penglai ship no. 1, these identifications

Table 6.6. Woods used in Penglai shipwrecks nos. 1, 2, 3, and 4

Hull component	Common name	Botanical name	Chinese Description
PENGLAI SHIP NO. 1			
Main keel	Pine	*Pinus* sp.	Song mu (松木)
Aft keel	Camphor	*Cinnamomum* sp.	Xiang mu (樟木)
Bulkhead	Castanopsis	*Castanopsis* sp.	Zhu li (锥栗)
Plank (Sample 1)	Chinese fir	*Cunninghamia lanceolata* (Lamb.) Hook.	Shan mu (杉木), sha mu (沙木)
Plank (Sample 2)	Chinese fir	*Cunninghamia lanceolata* (Lamb.) Hook.	Shan mu (杉木), sha mu (沙木)
Plank (Sample 3)	Chinese fir	*Cunninghamia lanceolata* (Lamb.) Hook.	Shan mu (杉木), sha mu (沙木)
Mainmast step	Phoebe	*Phoebe* sp.	Nan mu (楠木)
Stern	Phoebe	*Phoebe* sp.	Nan mu (楠木)
PENGLAI SHIP NO. 2			
Keel (main and aft)	Pine	*Pinus* sp.	Song mu(松木)
Garboard on starboard	Pine	*Pinus* sp.	Song mu(松木)
Hog piece	Elm	*Ulmus* sp.	Yu mu (榆属)
Forward keel	Elm	*Ulmus* sp.	Yu mu (榆属)
Bulkhead no. 4 second plank	Elm	*Ulmus* sp.	Yu mu (榆属)
Bulkheads (no. 1, second plank; no. 3 planks; no. 4, first plank)	Castanopsis	*Castanopsis* sp.	Zhu li (锥栗)
Frames (nos. 2, 4, 5)	Castanopsis	*Castanopsis* sp.	Zhu li (锥栗)
Mast step	Castanopsis	*Castanopsis* sp.	Zhu li (锥栗)
Mast longitudinal support (?) no. 1	Castanopsis	*Castanopsis* sp.	Zhu li (锥栗)
Mast longitudinal support (?) no. 2	Camphor	*Cinnamomum* sp.	Xiang mu (樟木)

(continued)

Table 6.6—*Continued*

Hull component	Common name	Botanical name	Chinese Description
Plank (strake no. 10 on port side, strake no. 9 on starboard, other scattered planks inside hull and beneath keel)	Chinese fir	*Cunninghamia lanceolata* (Lamb.) Hook.	Shan mu (杉木), sha mu (沙木)
PENGLAI SHIP NO. 3			
Bottom median planks	Pine	*Pinus* sp.	Song mu (松木)
Bulkheads (no. 3, first and second planks; no. 4, first and second planks; no. 5, first and second planks)	Pine	*Pinus* sp.	Song mu (松木)
Garboards, port side and starboard side	Pine	*Pinus* sp.	Song mu (松木)
Planks (strake nos. 3, 4, 6 on port side; nos. 2, 3 on starboard side)	Pine	*Pinus* sp.	Song mu (松木)
Frames (nos. 1, 2, 3)	Pine	*Pinus* sp.	Song mu (松木)
Mast step (forward and main)	Pine	*Pinus* sp.	Song mu (松木)
Transverse bars (run through bottom planks)	Oak	*Quercus acutissima*	Ma li (麻栎)
Loose tenon (for bulkheads)	Chinese chestnut	*Castanea mollissima*	Ban li (板栗)
Scattered timber (second plank from bottom of the hull)	Alder	*Alnus* sp.	Qi mu (桤木属)
PENGLAI SHIP NO. 4			
Unknown timber (4b1, 4b2, 4b3)	Pine	*Pinus* sp.	Song mu (松木)
OTHER			
Unknown timber (incomplete)	Ironwood?	*Erythrophleum* sp.	Ge mu shu (格木属)

demonstrate that the assemblage of the woods in the four ships represents ten taxa: *Pinus* sp., *Castanopsis* sp., *Ulmus* sp., *Cinnamomum* sp., *Cunninghamia lanceolata*, *Phoebe* sp., *Quercus acutissima*, *Castanea mollissima*, *Erythrophleum* sp., and *Alnus* sp. From the report, *Pinus* sp. appears to be the dominant wood. The main keel of Penglai ship no. 1 is made of pine, as are the main and aft keels of Penglai ship no. 2. Penglai ship no. 3, which shows characteristics of the Korean shipbuilding tradition, consists of multiple strakes instead of a keel, and the strakes are made of pine too.

Woods used for other important components of the hull of Penglai ship no. 2 vary but include *Ulmus* sp., *Castanopsis* sp., *Cinnamomum* sp., and *Cunninghamia lanceolata* for the hog piece, the forward keel, bulkheads, hull planks, frames, and mast steps. In some cases, the same component includes different woods. For example, both *Ulmus* sp. and *Castanopsis* sp. have been identified in samples from the bulkheads. In contrast, Penglai ship no. 3 shows consistency in the use of woods. Besides the bottom hull, all other hull components—the bulkheads, hull planks, frames, and mast steps—are also made of pine. A few other woods have been identified in wooden nails and dowels, including *Quercus acutissima*, *Castanea mollissima*, and *Alnus* sp. The differences in the wood assemblage between Penglai ships nos. 2 and 3 represent a technological difference between the two ships: Penglai ship no. 2 follows Chinese shipbuilding principles, whereas Penglai ship no. 3 shows Korean shipbuilding technology.

There is evidence that *Pinus* sp. and *Quercus* sp. are the dominant woods identified in Goryeo dynasty Korean ships (table 6.7). Pine has been identified in the bottom strakes of the hulls and in most of the hull planking. The beams, which secure the transverse strength of the hulls, are made from relatively hard woods, identified as *Quercus* sp., *Platycarya* sp., and *Morus* sp. Some of these woods are also used for the tenons and bars that lock the hull planks. Comparing the species identifications in the shipwrecks demonstrates consistency in the selection of particular woods. In the Korean shipbuilding tradition, wood has been used as a fastening material for a long time, and for the fastenings, hardwoods have been selected.

Use of Ironwoods in Shipbuilding

Wood remains identified as *Erythrophleum fordii* were found at the Ming dynasty shipyard in Nanjing; these show greater relative density values

Table 6.7. Woods used in Goryeo dynasty ships

Hull component	Common name	Botanical name
Wando ship		
Bottom plank (底板)	Pine	*Pinus* sp.
Bottom plank (底板)	Kaya (Japanese nutmeg-yew)	*Torreya nucifera* (L.) Siebold et Zucc.
Plank (外板)	Pine	*Pinus* sp.
Plank (外板)	Oak	*Quercus* sp. (*Quercus acutissima*?)
Beam A (加龍)	Oak	*Quercus* sp. (*Quercus acutissima*?)
L-shaped chine strake (彎曲縱通材)	Oak	*Quercus* sp.
L-shaped chine strake (彎曲縱通材)	Walnut	*Platycarya* sp. (*Platycarya strobilacea*?)
Free tenon (pisak, 皮槊)	Pine	*Pinus* sp.
Free tenon (pisak, 皮槊)	Zelkova	*Zelkova* sp.
Talido ship		
Bottom plank (底板)	Pine	*Pinus* sp.
Plank (外板)	Pine	*Pinus* sp.
Beam A (加龍)	Oak	*Quercus* sp. (*Quercus acutissima*?)
Beam B (駕木)	Oak	*Quercus* sp.
Beam C (駕木型加龍)	Mulberry	*Morus* sp.
Free tenon (pisak, 皮槊)	Mulberry	*Morus* sp.
Sibidongpado ship		
Bottom plank (底板)	Pine	*Pinus* sp.
Plank (外板)	Pine	*Pinus* sp.
Beam A (加龍)	Walnut	*Platycarya* sp. (*Platycarya strobilacea*?)
L-shaped chine strake (彎曲縱通材)	Walnut	*Platycarya* sp. (*Platycarya strobilacea*?)
L-shaped chine strake (彎曲縱通材)	Pine	*Pinus* sp.
Bow plank (船首板材)	Pine	*Pinus* sp.
Free tenon (pisak, 皮槊)	Walnut	*Platycarya* sp. (*Platycarya strobilacea*?)
Tenon (changsak, 長槊)	Oak	*Quercus* sp. (*Quercus acutissima*?)

(*continued*)

Hull component	Common name	Botanical name
Anjwa ship		
Bottom plank (底板)	Pine	*Pinus* sp.
Plank (外板)	Pine	*Pinus* sp.
Beam A (加龍)	Oak	*Quercus* sp. (*Quercus acutissima*?)
Beam C (駕木型加龍)	Pine	*Pinus* sp.
Stern (船尾板材)	Pine	*Pinus* sp.
Rudder (舵推定)	Oak	*Quercus* sp. (*Quercus acutissima*?)
Free tenon (pisak, 皮槊)	Pine	*Pinus* sp.
Tenon (changsak, 長槊)	Pine	*Pinus* sp.
Ansan ship		
Bottom plank (底板)	Pine	*Pinus* sp.
Plank (外板)	Pine	*Pinus* sp.
Free tenon (pisak, 皮槊)	Oak	*Quercus* sp. (*Quercus acutissima*?)

(0.90–1.06). However, heavier woods with a relative density of more than 1.00 have not been identified among excavated ship remains. Extremely heavy woods have common features, including thick and lignified cell walls and significant concentration of their fiber, and the heartwood of these woods is very dark. Because of their weight, hardness of structure, and color, they are commonly called ironwoods. The term *ironwood* is used for various heavy and hard woods in many places. Ken Ogata (1972) listed approximately ninety woods recognized as ironwoods. A family group having many ironwoods in the lower taxa is Leguminosae (the bean family). *Erythrophleum fordii* is in this family. Heavy and hard woods are also identified as belonging to the families Myrtaceae, Sapotaceae, and Dipterocarpaceae, mostly growing in the tropical regions.

Heavy and hard woods have been used as construction materials, while lightness and softness are important in terms of wood's utility. The weight of timber also has implications in terms of transportation costs, which are more significant for heavy and hard woods than for light woods in many cases. Nevertheless, the usefulness of the heavy and hard woods has been understood for a long time in China. The historical text *Lingwai Daida*, written in 1178, describes large foreign vessels using a very hard wood for their rudders (Shiba 1968: 224; Needham 1971: 645).

> The Chinchow coastal mountains have strange woods, of which there are two remarkable kinds. One is the tzu-ching-mu (purple thorn tree) as hard as iron and stone, in colour red as cosmetic paint, and straight-grained; as large in girth as two men's reach, and when used for roof beams will last centuries. (*Lingwai Daida*, vol. 6, in Needham 1971: 645)
>
> 欽州海山有奇材二種．一曰紫荆木，堅類鐵石，色比燕脂，易直合抱以為棟梁，可數百年．一曰烏婪木，用以為大船之柂，極天下之妙也．（嶺外代答 巻六）

According to the *Lingwai Daida*, two kinds of special trees originated in coastal regions and the mountains of Qinzhou, which is located in Guangxi Province close to Vietnam. Besides its color, the tree known as *zu ching mu* (紫荆木), identified as *Cercis* sp. (Needham 1971: 645), is characterized by being "as hard as iron." The historical text mentions another hardwood tree known as *wu lan mu* (烏婪木) that was used for the rudders of large ships and was regarded as the finest wood for that purpose. However, details about this tree, including its species, have not been clarified. The text describes the tree in terms of its great density and durability, which were necessary for large seagoing vessels to sail long distances in the deep southern seas. Foreign ships from Southeast Asia sailing on the South China Sea seem to have had their steering oars or rudder made of wood from this tree. The description in the *Lingwai Daida* mentions the value of the hardwoods and the matter of supply: when purchasing the wood, people in Guangdong and Wenzhou must have paid ten times as much as in Qinzhou, because the supply of this wood was only 10 or 20 percent of the demand in these areas as a result of problems transporting the wood by sea (Needham 1971: 645). The description thus provides evidence for the limited distribution of hardwoods from southern regions, which would have increased their value, and further indicates the existence of timber markets and a distribution system for shipbuilding industries in China in the twelfth century.

Timber Industries and Shipbuilding Industries

The availability of ship construction materials during the tenth to thirteenth centuries can be understood from a review of Chinese industries at the time. During the Song dynasty, ship transportation industries owned by governmental and nongovernmental agents were growing and contributed to the widespread supply of ship construction materials to shipyards in China (Shiba 1968: 222). Along with substantial growth in various other industries, new ship transportation facilities appear to have formed, and they started to occupy hinterlands of rural areas.

These phenomena did not occur all over the Chinese continent: in general, the timber industry in the north was not as productive as in the south. Shiba (1968: 223) cites a description in the twelfth-century text *Zhong Mu Ji* (Collection of Zhong mu, 忠穆集) that mentions the superiority of woods from the southern part of China. The seagoing ships from Fujian were judged to be the finest, and those from Guangdong and Guangxi were second, with those from Wenzhou and Mingzhou (Ningbo) following as third. Wood growing in the north was considered inadequate for ships sailing on open seas. Moreover, a paucity of environments supplying timber in northern China can be inferred from descriptions in historical records, which mention that the construction of cargo ships ceased on the upper Yangtze River because of an insufficient supply of appropriate timber (Shiba 1968: 222–23).

The common use of the camphor tree, Chinese pine, and Chinese fir demonstrated in the archaeological data corresponds with historical research. Some geographical names and shipyards using the wood of these trees appear along with tree characteristics in historical botanical texts, such as *Zhenglei Bencao* (Classified materia medica, 証類本草). In the shipyards of the Jiangnan regions around the lower Yangtze River, wood from the camphor tree has served as the most durable ship construction material because of its hardness.

Another source provides a record of the construction of 168 ships using pine wood during the reign of Emperor Xiaozong (1162–89) around what is now Hunan Province (Shiba 1968: 223). Shiba (1968: 223) mentions a description in a historical botanical text indicating that Chinese fir was commonly used because of its resistance to water damage. According to Shiba (1968: 224), some descriptions quoted in Ming dynasty texts, including *Ming Shi* (History of Ming, 明史) and *Wanli Huidan* (Collected statutes of Wanli, 万曆会典), cite the quality of timbers used for ships: cedar is the most suitable and expensive wood, and wood from the camphor tree and pine follow, but the quality of pine wood cannot match the others. Woods ideographically described in the texts need to be identified botanically before we can assess their correspondence with wood remains from archaeological sites; that is beyond the scope of this research.

Jiangxi Province has been a focal area because of its conjunction of inland shipping and timber supply. The rise of shipyards in this area coincided with the development of forestry. The area could have played a significant role as a source of ship timbers being supplied to coastal areas through river transportation and portage systems. The establishment of shipyards on the coasts was substantially related to supplies from the hinterland. Wenzhou in southeastern Zhejiang Province and Mingzhou (now the city of Ningbo) both flourished as centers of shipbuilding. Mingzhou, however, could have been the more productive, relying on adjacent Chuzhou (Lishui), which was famous for its timber industry. Wenzhou and Mingzhou were both focal areas for the collection and distribution of timber for export. Traders from Guangnan and Fujian and Japanese merchants visited the two cities to trade in seaborne timber. Government shipyards were established in these two cities by the time of the reign of Emperor Renzong (1022–63). Compared to inland shipyards along rivers, the shipyards in Wenzhou and Mingzhou could have been growing independently. This was because they were major centers dealing in ship construction materials, including not only timbers but also other essential products, such as oil, iron products, lime, vegetable fiber, and coal. In the end, the shipyards in Wenzhou undertook construction of both seagoing ships and river ships, following government orders. However, the decline of the shipyards eventually occurred, because of exhaustion of the wood resources. Mingzhou would retain dominance as an international port in the following centuries.

Perspectives on Ship Construction Materials

Reviewing ship construction materials and related industries leads to a better understanding of how shipbuilding technologies developed in the East China Sea regions. Fastening methods using iron materials were common in the East China Sea shipbuilding tradition. Early innovations in iron production in China facilitated the use of iron for ship construction, and iron was widely used for nails, clamps, and brackets during the Song and Yuan dynasties—these are the three major types of iron fastenings used in the period under consideration.

The three major types of trees likely to be identified in excavated ships from mid- to southern China are Chinese pine, *Pinus massoniana*; Chinese fir, *Cunninghamia lanceolata*; and the camphor tree, *Cinnamomum camphora*. Selecting the appropriate woods for ship construction was done with concern for their utility and availability. Utility is assessed above in terms of timber utilization as a ship construction material, with a focus on characteristics of the wood, such as hardness and resistance to degradation. Availability is assessed taking into consideration the extent of the timber industry during the Song dynasty.

7

East China Sea Rising

The rise of the East China Sea shipbuilding tradition in China during the Song dynasty facilitated Chinese merchants' seafaring by providing seagoing ships built in their own country. Substantial changes in the governing system in mainland China during the period from the Song dynasty (AD 960–1279) to the Yuan dynasty (AD 1271–1368) impacted maritime activities widely in East Asia, yet maritime networks and human movement by shipping expanded throughout these periods (Yasuhiro 2008), and technological innovations in shipbuilding continued to occur. Excavated ships in China dating from the Yuan and Ming dynasties indicate the existence of still another shipbuilding tradition or subtradition in the East China Sea, distinct from those described in the preceding chapters.

The Song and Yuan Dynasties

Socioeconomic growth and the expansion of trading networks by shipping were more apparent during the Song dynasty in China than in previous periods (Hartwell 1966; Shiba 1968; Clark 1991; Chaffee 2010), partly because of further development of the country's water transportation system: "Progress of water transportation showed itself in general in the construction, propulsion and control of ships (both on the ocean and in the inland waterways) as well as in the growing specialization of ships according to their purpose, their region of operation and the like" (Shiba 1968: 7 [my translation]). Such specialization occurring in the water transportation industries coincided with the growth of ship manufacturing; many shipyards for inland watercraft developed in Wenzhou and Mingzhou. Around the later period when the Quanzhou ship was built, the Song dynasty had already lost parts of its territory to invasion by Tungusic groups from the north. The dynasty had managed to continue its reign by relocating its capital from Beijing to the southern city known as Linan (now

Hangzhou), whereupon it became known as the Southern Song dynasty. After this drastic change, it still administered a water transportation system by relying on shipbuilding industries in the south, and it maintained naval power until the full-blown rise of the Yuan.

During the Song dynasty, the private and governmental sectors were both widely involved in water transportation and related industries, resulting in a diverse suite of ship owners, including rich farmers, lords, merchants, Buddhist clergy, and regional and central bureaucrats (Shiba 1968). Industrialization of the shipyards, in particular, was under government control. Interference with the private shipyards of the south through governmental control is presumed to have occurred after the southward relocation of the capital. The consequences of interference may be related to changes in the shipbuilding traditions in the central and southern parts of China.

Song Dynasty Seagoing Ships

China has seen an increase in archaeological discovery of coastal and seagoing ships dating to the thirteenth century and constructed in regions south of Hangzhou Bay. As discussed in previous chapters, technological innovations occurred in parts of the zone from Zhejiang southward; these innovations include the combination of bulkheads and a keel to produce a type of hull better suited to seafaring. Bulkheads were the main transverse components in coastal or seagoing traders. Other features developed in the hull structure that enhanced its sturdiness for seagoing voyages include the assemblage of frames and brackets attached to either side of the bulkheads to affix them to hull planking. The hull planking of the coastal or seagoing ships was often characterized by some sort of clinker construction method. Adoption of sacrificial planks also became a common feature.

Aside from the Quanzhou ship, hull remains of the Song dynasty's seagoing ships discovered in China include Bai Jiao shipwreck no. 1, Nanhai shipwreck no. 1, and Huaguang Reef shipwreck no. 1 (Kimura 2010a). These ship remains are thought to date back to the late thirteenth century and to have originated from the East China Sea region. Only a few timber remains survived at the Bai Jiao shipwreck no. 1 site (Kenderdine 1995, 1997). Nanhai shipwreck no. 1 and Huaguang Reef shipwreck no. 1 have substantial potential to be important resources comparable with the

Quanzhou ship and Shinan shipwreck. However, only limited information is currently available about the hull remains.

Nanhai Shipwreck No. 1 (南海一号沉船)

Nanhai shipwreck no. 1 in the South China Sea is the best-known shipwreck in terms of attracting public interest, not only because of its historical significance but also because of the state of preservation of its hull and cargo. Details of this shipwreck are available in a few sources (Zhang 2006a; Liu 2011; Wei 2011; Yiran 2014). The shipwreck was accidentally discovered in 1987 by the Guangzhou Salvage Bureau under the Ministry of Communications and a British salvage company, while they were searching for Dutch East India Company shipwrecks offshore near Shangchuan and Xiachuan Islands. The significance of the site was recognized after some precious artifacts were recovered, and immediately after that the site was protected.

In 1989 the shipwreck site was initially inspected by an international team from the National Museum of Chinese History and an avocational organization for underwater archaeological survey from Japan. Beginning in 2001, a Chinese team from the Underwater Archaeological Research Center of the National Museum commenced a series of surveys and partial excavations. As a result of that exploration, which continued until 2004, a challenging approach was adopted, aimed at raising the ship remains along with the surrounding sediment for the sake of presenting underwater archaeological excavation work on the shipwreck within a caisson inside a museum. In 2007 a large caisson was placed around Nanhai shipwreck no. 1; it was then recovered containing both the shipwreck and its physical seabed context. It was moved to the newly built museum on Hailing Island in Yangjiang city, Guangdong. Almost two decades after the discovery, the entire shipwreck, still embedded in the sediments, was placed on land.

In the following years, Chinese researchers continued removing the sediment and exposing the hull in the museum. The results of the partial excavation and the prior remote sensing survey indicate that the shipwreck remains in the caisson are approximately thirty meters long, seven meters wide, and four meters deep. At this stage, detailed information about the hull is not available. Numerous recovered ceramics originating from Jingdezhen, Jiangxi, Longquan, Zhejiang, Dehua, and Cizao indicate that the ship was a trader. Many copper coins were also recovered;

these mainly date to the Northern Song dynasty. Among them, however, are two coins inscribed as Jianyan Tongbao (建炎通寶) and Shaoxing Yuanbao (紹興元寶) that date the wreck to the Southern Song dynasty. Relevant academic reports have not yet been published, and only partial information about the recovered artifacts is available (Tian et al. 2014).

Huaguang Reef Shipwreck No. 1 (華光礁一号沉船)

A Chinese underwater archaeology team conducted underwater archaeological surveys and excavations in the Paracel Islands (Xisha Islands) in the late 1990s. During the seasonal work, thirteen sites were identified, dating from between the period of the Five Dynasties (AD 907–60) and the twentieth century (Zhang 2006b). They include one well-preserved shipwreck. The remaining part of the hull was found in 1996 at the south-central reef known as Discovery or Huaguang Reef, causing it to be named Huaguang Reef shipwreck no. 1. Immediately after the hull was located, the site was disturbed by looters.

Through 1998 and 1999, the Chinese team conducted surveys and test excavations. More than 850 pieces of ceramics, ship timbers, and other artifacts were recovered. In 2007 a full excavation commenced, and the preservation status of the hull and site were assessed. More than 6,000 artifacts were recovered in this operation, including iron ingots and many ceramics (including some intact wares). During the second season of excavation, 511 ship timbers were recovered. The recovered timbers were taken to a museum on Hainan Island in Hainan Province for preservation and analysis. An artifact report with a focus on ceramic study has been published (Shanxi Museum and Hainan Provincial Museum 2013), but detailed information about this shipwreck hull is not yet available to researchers outside the country. A Chinese government officer provided a brief explanation of the site at international conferences in 2009 (Yang 2009).

According to Zelin Yang of the Institute of Cultural Relics and Archaeology of Fujiang Museum, the remaining hull measures approximately seventeen meters in length and seven and a half meters in width. Its state of preservation is very good. Three members of the keel remain. The hull planking shows five layers in some parts. The first and second inner planks are thicker than the outer planks. Iron nails were used for the hull planking, but details about these have not been reported. There are ten bulkheads: the seams of the bulkhead planks show rabbeted joints, and there are limber holes in the bulkheads. The manner of joining the bulkheads to the hull planks is similar to that of the Shinan shipwreck in the use of wooden brackets and frames (Yang 2009). The shipwreck has been dated to the Song dynasty. Some blue and white porcelain pieces were found in the hull, and some ceramics were discovered beneath the shipwreck. A large metal concretion was in the middle of the hull.

Ningbo Ship (寧波船)

Among the Song dynasty excavated seagoing ships in China, one shows a structure that is different from that of the others. Discovery of the Ningbo ship dates back to 1979, similar to the Quanzhou ship. The hull of the Quanzhou ship was secured for exhibition; however, the fate of the Ningbo ship was not as favorable. Though the remaining part of the hull was documented and we can consult the description in both Chinese and English, the hull itself is no longer extant (Lin et al. 1991; Green 1997a). Drawings of it indicate differences from other ships in the composition of the keel and garboards, which do not provide deadrise steep enough to form a V-shaped midship cross section, and the cross section also shows that the hull planking is not in a single layer (figure 7.1).

The archaeological remains of the Ningbo ship were discovered near the Fenghua River (奉化江) at Dongmenkou in the city of Ningbo (Lin et al. 1991). After the hull was exposed, deterioration occurred quickly, and unfortunately the hull was not properly preserved. The surviving portions of the hull (measuring 9.3 meters in length and 4.4 meters in width) consist of the forward section of the hull bottom, including the forward keel, the main keel, bulkheads, and hull planking. The remaining portion of the main keel is 7.34 meters long, 0.26 meters wide, and 0.18 meters deep. The middle member of the main keel shows the greatest length (5.10 meters), and both ends are stepped for the scarfs. On its aft scarf, a piece of timber remains that is regarded as part of the aft member of the main keel or an aft keel. Since this is unclear, the number of original members making up the main keel has not been ascertained.

The forward member of the main keel is approximately 2 meters long, and its forward end is scarfed to join the forward keel. The forward keel angles upward at 35 degrees. Its remaining part is 1.55 meters long and has a triangular cross section, the widest part measuring 180 millimeters with a depth of 200 millimeters. There are twelve coins and associated small holes in the scarf joint between the main keel and the forward keel,

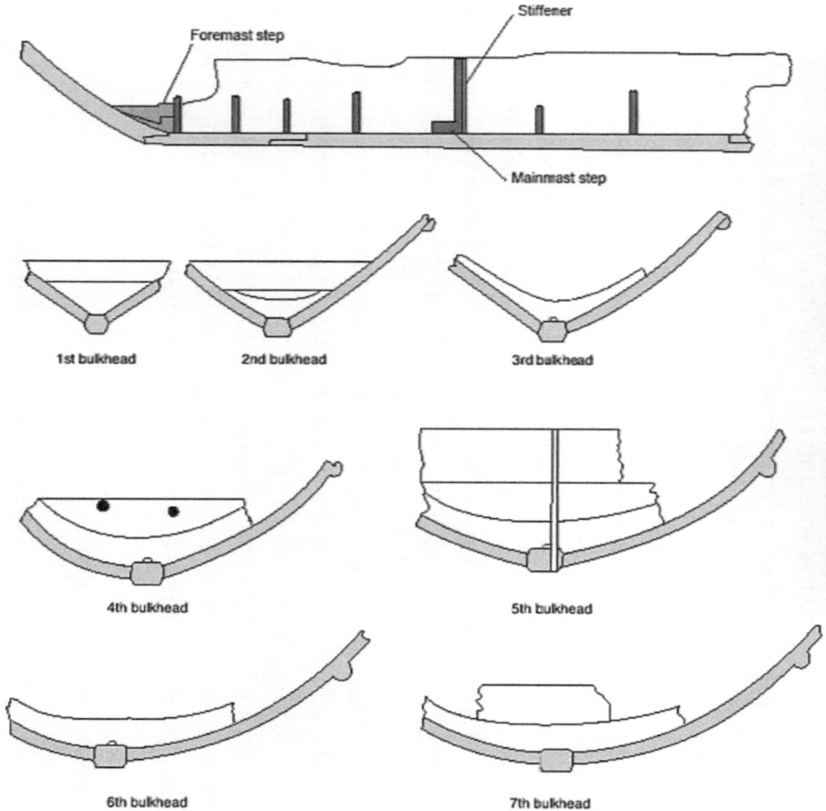

Figure 7.1. Ningbo ship, plan, cross sections, and photograph of the exposed hull. (From Green 1997)

associated with the previously discussed shipwrights' tradition (Green 1997a: 20).

Seven bulkheads remain on the main keel. The thickness of the bulkhead planks is 70–100 millimeters, on average. The bulkheads are nailed to the frames, which in turn are nailed to the hull (Green 1997a: 20). A supporting timber or stiffener is vertically installed against the aft side of the fifth bulkhead, and this timber is fixed into a recess on the keel. Because this arrangement appears to be only on the fifth bulkhead, which is the position of a mast step, the timber is presumably for reinforcing the bulkhead against the loads imposed by the mast. Bulkhead spacing varies; the smallest hold is the second, measuring 620 millimeters between bulkheads, as compared to the fifth hold, measuring 2.05 meters between bulkheads, which is the largest space. A forward mast step is placed on the forward side of the first bulkhead, and it measures 840 millimeters athwartships, 210 millimeters fore and aft, and 140 millimeters thick. A main mast step, placed in the forward side of the fifth bulkhead, measures 1.04 meters athwartships, 250 millimeters fore and aft, and 180 millimeters thick. Both mast steps have recesses to receive tabernacle cheeks.

The cross-section drawings show a gentle turn to the bilge starting at the garboards, with a moderate deadrise. In the bow, the cross-section drawing shows a sharp V shape. The hull planking is carvel-built, and according to the drawings, it is single layered. The length of the remaining plank is 3.0–8.0 meters, with a width of 210–420 millimeters and a thickness of 60–80 millimeters. Iron nails having a rectangular section measuring 10–15 millimeters on each side are used for the hull planking, and these nails have been skew driven at intervals of 100–250 millimeters. The seams of the strakes are filled with putty. On the outer surface around the seventh and eighth strakes on the starboard side, a longitudinal timber with a semicircular profile is fixed by iron nails. It is 7.10 meters long and runs along the hull. This has been explained to function as a bilge strake to contribute to stability and strength (Wang 2000: 144). In addition, what is thought to be part of the rudder stock has remained.

Yuan Dynasty Seagoing Ships

The establishment of the Yuan dynasty, ensuing from the Mongolian occupation of China, substantially affected relationships among nations and changed maritime networks between China and the other East Asian countries, including the destruction of the tribute system. The newly restricted relationships between countries allowed private traders to make more inroads into overseas trading instead of the seaborne tribute trading that had previously occurred. Yet the Mongolian rulers of the Yuan dynasty appreciated the significance of water transportation and beneficial maritime trade as much as previous dynasties had (Matsuura 1995: 39). Based on that understanding, their policy involved strictly controlling piracy yet being conciliatory toward illegal traders. Chinese ships were employed as a means of overseas transportation. The significance of the ceramics trade in the northern part of China is indicated by the recovery of many ceramics from the Sandaogang wreck site offshore of Suizhong in Huludao city, Liaoning Province, by a Chinese underwater archaeological team (Zhang 2001; Kimura 2010a).

A few ship remains dating to the Yuan dynasty and the following Ming dynasty (AD 1368–1644) show defensive features: these ships carried weapons, and some of the recovered ships were initially designed as battleships and patrol boats. One of the best-preserved Chinese ships, from the city of Penglai in Shandong Province, has been recognized as a battleship or patrol boat. Armed ships or ships carrying firearms appear to have become more numerous during the late Yuan dynasty period. There are probably multiple reasons for why the ships were armed. One factor relates to the necessity of self-defense among private traders, who needed to protect their investments from piracy; another could involve an increase in smuggling, which became more serious as the Yuan dynasty declined.

The Penglai Ancient Ships

Penglai is a northern city facing the Yellow Sea—well to the north of where the Song dynasty seagoing ships were found. Underscoring the significance of the Penglai discoveries is that they show structures very different from those of excavated ships from the south. The presence of a keel, for example, is not clear in the hull structure, whereas use of a keel is identified in representative seagoing vessels such as the Quanzhou and Shinan ships. Other differences include the form of the hull, which does not show a greater width than the Song dynasty's ships; distinctive fastening methods for the hull planking; and no use of sheathing planks (the hull planking consists of a single layer). Notably, these features are identical to those of the Ningbo ship.

Penglai Ship No. 1 (蓬莱古船一号)

Penglai ship no. 1 was the first to be discovered among the four sets of ship remains known as the Penglai ancient ships. They were found near Penglai Castle at Yantai city in Shandong Province. In 1984, during dredging at Dengzhou Harbor, the first shipwreck was encountered. It has been documented in various sources (Xi 1989; Xi and Xin 1991; McGrail 2001: 371–73; Blot 2006; Yuan 2006b).

The remaining part of the hull is 28.6 meters long and 5.6 meters wide. There is a narrow flat bottom and then a gentle turn to the bilge (figure 7.2). The shape in cross section above that turn of the bilge appears to be very flared. The well-preserved lower hull includes a keel and thirteen bulkheads. The keel consists of three members joined by hooked scarf joints: the forward keel is 3.60 meters long; the main keel, 17.06 meters; and the aft keel, 5.58 meters. The main keel shows slight hogging. There are recesses for placing mirrors or coins in the keel joints. The keel scarfs are reinforced by hog pieces fitted on the top of the keel.

The best-preserved bulkheads, no. 3 and no. 5, consist of four planks respectively and are 160 millimeters thick, on average. Part of the upper surface of the three lower planks is rabbeted to form a tongue and groove joint, and the uppermost bulkhead planks have four holes that could be used to place longitudinal timbers through the hull. Each bulkhead plank is joined by four sets of mortises and tenons, and iron clamps appear to fasten the lower and upper bulkhead planks. There are two limber holes in each bulkhead, about half a meter from the keel on either side. The bulkheads located forward show that half frames are attached to their aft side and iron brackets are used to fix the bulkhead planks in the hull planking on the forward side. The reverse arrangement pertains on the bulkheads located aft of midships: they have half frames on the forward side and L-shaped iron brackets on the aft side.

Ten strakes remain on the port side, and eleven strakes remain on the starboard side. The hull planks are 3.7–18.5 meters long, 200–440 millimeters wide, and 120–280 millimeters thick. Garboard strakes show the greatest thickness; they are baulks rather than planks. The hull planking is edge-joined by two different types of square iron nails: large edge-driven nails and bent skewed nails. The hull planks are joined to make up strakes with hooked scarf joints having mortise-and-tenon ends.

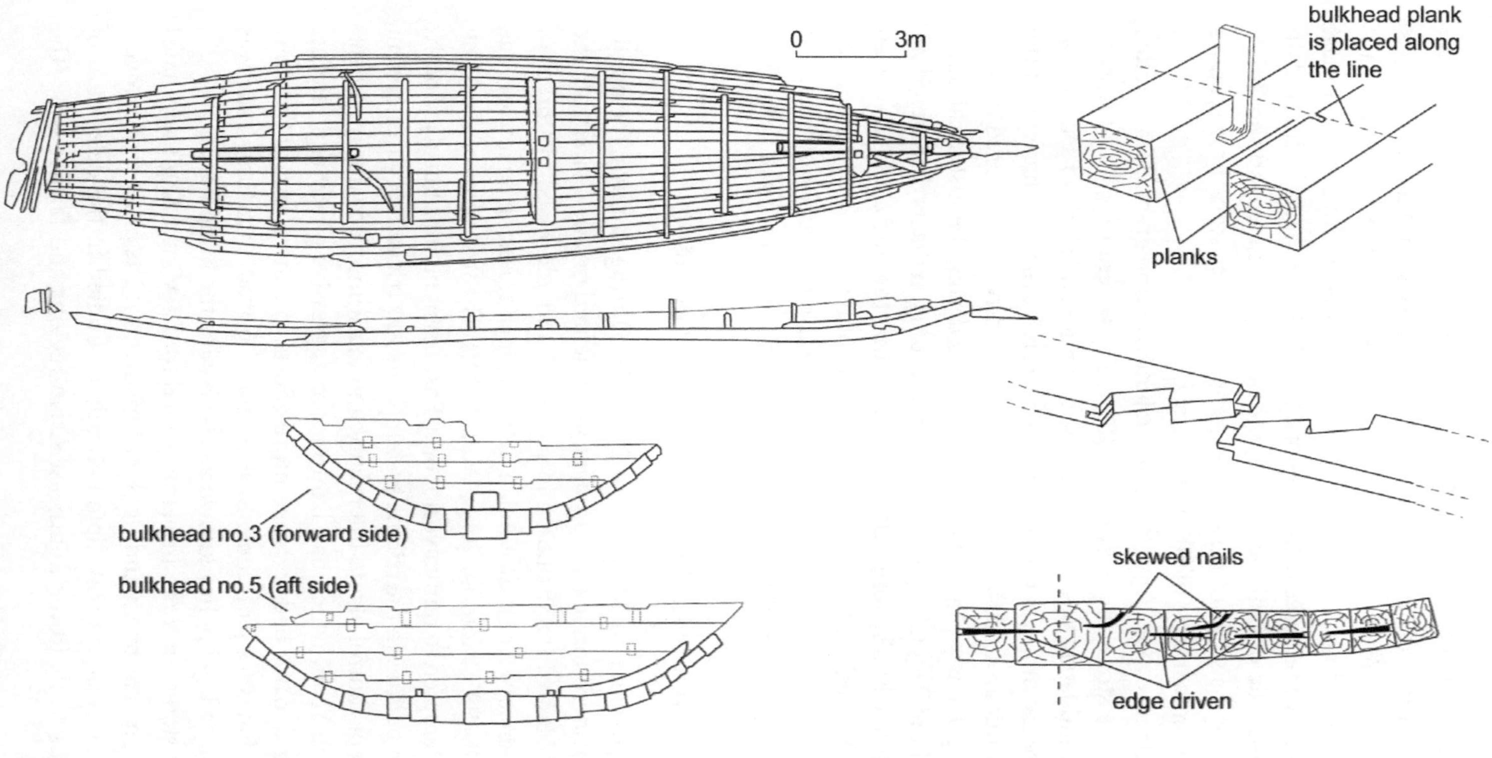

Figure 7.2. Penglai ship no. 1, plan and diagrams showing the use of metal brackets to secure hull planks and a bulkhead plank, a hooked scarf for the hull planks, and two different kinds of nails (skewed and edge-driven nails) fastening the hull planks. (From Xi 1989 and Cultural Relics and Archaeological Institute of Shandong Province et al. 2006)

Propulsion of the ship is evidenced by two mast steps remaining on the hull. A forward mast step having two recesses with a size of 200 × 200 millimeters to receive tabernacle cheeks is located on the forward side of bulkhead no. 2, measuring 1.6 meters athwartships. A mainmast step having two recesses with a size of 260 × 260 millimeters is located on the forward side of bulkhead no. 7, measuring 2.9 meters athwartships. A transom consisting of three timbers has a rudder stock hole (rudder trunk) with a diameter of 300 millimeters.

Artifacts recovered from the ship include five iron grapnels, a wooden anchor, three stone anchors, ropes, 489 Chinese ceramics, a few Korean ceramics, and Japanese coins. From the ceramics and study of sedimentation where the hull was discovered, the research team concluded that the ship was built around the end of the Yuan dynasty and was used into the early Ming dynasty (Yang 1989: 61–62). Considering the historical background of the Penglai Castle and a few weaponry artifacts discovered inside the hull, Chinese researchers speculate that the ship could have been used as a battleship; they refer to a ship mentioned as *daoyu zhanzhao* (sword-fish battleship, 刀鱼战棹) and to the Qing dynasty historical text *Penglai Xianzhi* (History of Penglai, 蓬萊縣志 [Zou and Yuan 1989: 76]).

Rise and Decline of the Ming Dynasty (AD 1368–1644)

The Ming dynasty ruled China for 270 years. During its reign, the political and socioeconomic environment was oriented in a new direction: the revival of Han China (as noted in chapter 3 and addressed in studies of the emergence of a new East Asian network encompassing Southeast Asia [Reid 1988, 1995; Momoki 2008; Wade 2010]). The first ruler of the Ming empire attempted to reinvolve not only East Asian nations but also Southeast Asian polities in the tribute system. This was triggered by insufficient government control of and intervention in maritime trade during the decline of the Yuan, which had led to more-aggressive piracy and raiding. The Ming dynasty had to make more efforts to control piracy than any previous dynasties had done. Restriction of private maritime trade was enforced by a legal policy known as Hai Ban or the Ming Ban.

Inconsistency in Yuan imperialism led to the Ming bringing in naval power as a means of expanding their jurisdiction. The process of recovering hegemony at sea was diplomatically embodied in the delegation of Zheng He's fleets to the Southeast Asian regions and farther west. The purpose of Zheng He's voyages was to demonstrate to those nations, either implicitly or explicitly, the reassertion of a Han Chinese empire (Needham 1971: 487–88; Zheng 1991). Non-Chinese scholars have interpreted this as a form of armed diplomacy (Matsuura 1995: 46; Wade 2004; Danjo 2005: 160). The effects of Ming southern expansion through maritime activities have been pointed out in terms of the direct and massive dissemination of the Han Chinese system, rules, and dogmas into the kingdoms and nations of Southeast Asia (Laichen 2010). The formation of sea-lanes connecting various parts of Southeast Asia is more clearly recognizable in the Ming period than in previous eras (Mills 1979). John Miksic (2010) has addressed the issue archaeologically, reviewing how the Ming voyages and maritime policies affected Southeast Asian trading patterns.

The construction of seagoing traders during the Ming dynasty is hard to evaluate. Archaeological evidence of ships that might have been engaged in long-distance voyages to Southeast Asian and Indian Ocean regions is limited in Chinese waters. One exception to this general statement is the Nan'ao shipwreck, for which the discovery of numerous blue and white ceramic pieces suggests that it was engaged in exporting ceramics outside China (Jian 2011). The report published in 2014 provides basic knowledge about the ceramic assemblage, classified into eleven different types (Guangdongsheng Wenwu Kaogu Yanjiusuo 2014), but the information about the hull remains is limited. There are more sources that discuss the remnants of other coastal ships found in China. Technologies evident in the remains of coastal ships are nevertheless sufficient for developing some premises regarding the formulation of shipbuilding technologies in this period.

Ming Dynasty Trader: Nan'ao Shipwreck No. 1 (南澳一号)

The Nan'ao shipwreck was discovered by a local fisherman in 2007 off the Nan'ao Islands near Shantou city in Guangdong Province. The area where the ship was found is adjacent to Fujian Province and the Taiwan Strait. According to a brief report (Jian 2011), the Guangdong Archaeology Cultural Relic Research Institute, Guangdong Museum, and National Conservation Center for Underwater Cultural Heritage conducted a joint underwater archaeological excavation between April and July 2010. The excavation exposed an area of ten meters by thirty meters and revealed most of the hull remains. The length of the remaining part of the hull is approximately twenty-seven meters, and the maximum width is seven and a half meters. The hull was divided by fifteen bulkheads, producing sixteen compartments. While the condition of the hull is reported as better than that of other shipwrecks found in China, little detailed information about hull structure is available as yet. During the first season's excavation, 11,248 artifacts were recovered from the site; the majority, 10,624 pieces, are ceramics. Most are various types of blue and white wares, including plates, bowls, bottles, jars, and cups from the Zhangzhou kiln in Fujian Province. Some Jingdezhen wares were also recovered from the shipwreck. Apart from the blue and white ceramics, recovered items include wooden wares, iron pans, glazed jars, organic remains, copper coins and sheets, and miscellaneous copper products. The total number of recovered artifacts is over 54,000 pieces (Guangdongsheng Wenwu Kaogu Yanjiusuo 2014). Chinese researchers have concluded that the ship probably sailed from Fujian, which functioned mainly as an export port at the time (Jian 2011: 296), and thus should be considered a Ming dynasty trader.

Coastal Ships

Four additional Ming dynasty excavated ships have been archaeologically investigated: the Xiangshan Ming dynasty shipwreck and Penglai ships nos. 2, 3, and 4. The well-preserved hulls of the Xiangshan Ming dynasty shipwreck and Penglai ship no. 2 are important testimony by which to identify a cluster of shipbuilding technologies in the East China Sea shipbuilding tradition, in combination with the remnants of the Ningbo ship and Penglai ship no. 1.

After the discovery of Penglai ship no. 1 in 1984, another three ships were found in 2005 during development work adjacent to Penglai Castle. The importance of the discovery lies in the apparent structural differences between the hulls; Penglai ship no. 2 shows evidence of the Chinese shipbuilding tradition, while the other two ships reveal evidence of the Korean shipbuilding tradition in their structure and construction methods (as can be seen in cooperative research by Chinese and Korean scholars [Gong and Xi 2006; Yuan 2006a; Cui and Kim 2007; Cui 2009]). The adoption of Korean shipbuilding technology is clearly evident in the well-preserved Penglai ship no. 3, but that ship also has a bulkhead structure as in Chinese shipbuilding. Although the idea of hybridization has not been introduced or appreciated among scholars so far, what can be observed on Penglai ship no. 3 is that the hull represents technological integration in northern East Asia.

Xiangshan Ming Dynasty Shipwreck (象山明代沉船)

In 1994, the remains of a Ming dynasty ship were found in the silt of an ancient harbor near Qibu village in Xiangshan, Zhejiang Province. Beginning in 1995, the Archaeological Research Institute of Ningbo conducted an excavation of the ship (figure 7.3). The remaining part of the hull measures 23.7 meters in length and 4.9 meters in width and includes a keel and bulkheads (Xi et al. 2004: 195–98). The cross section shows a rounded bottom with sharper sections at the bow. Details of the keel have not been reported, but the keel seems to consist of three timbers. On each posited joint in the keel, there is a hog piece like those observed on Penglai ship no. 1.

Twelve bulkheads divide the hull, and two or three broad planks remain in each bulkhead. Some of the second bulkhead planks have two square holes measuring 18–20 × 14–20 millimeters to place two timbers running longitudinally through the hull. The bulkhead planks in the bottom have two limber holes. There are frames attached to the bulkheads. The distribution of the frames has not been clearly explained, but the drawings suggest that most bulkheads have frames on both the forward side and the aft side. The frames attached to the bulkheads seem to have been fastened to the hull planking by iron nails. Numerous hull planks remain, evidencing seventeen strakes per side through the midsection of the hull. The planks vary from 140 to 160 millimeters in thickness and from 80 to 200 millimeters in width and thus, because of their thickness, are technically baulks rather than planks. The hull planking is edge-joined and fastened by iron nails, though the details of the iron nails and the pattern of fastening are unclear. The seams of the hull planking were caulked with hemp and secured by a putty of tung oil and lime.

The hull planking extends beyond the aft transom, where part of the rudder remains. A forward mast step remains on the forward side of bulkhead no. 2, and a mainmast step remains on the forward side of bulkhead no. 6. In the bottom of the hull, besides bricks and roofing and ridge tiles, a few ballast stones were found. The date of the ship was determined from ceramics, including a porcelain bottle with a small mouth dating to the Yuan dynasty and a few celadon pieces from the Longquan kiln dating to

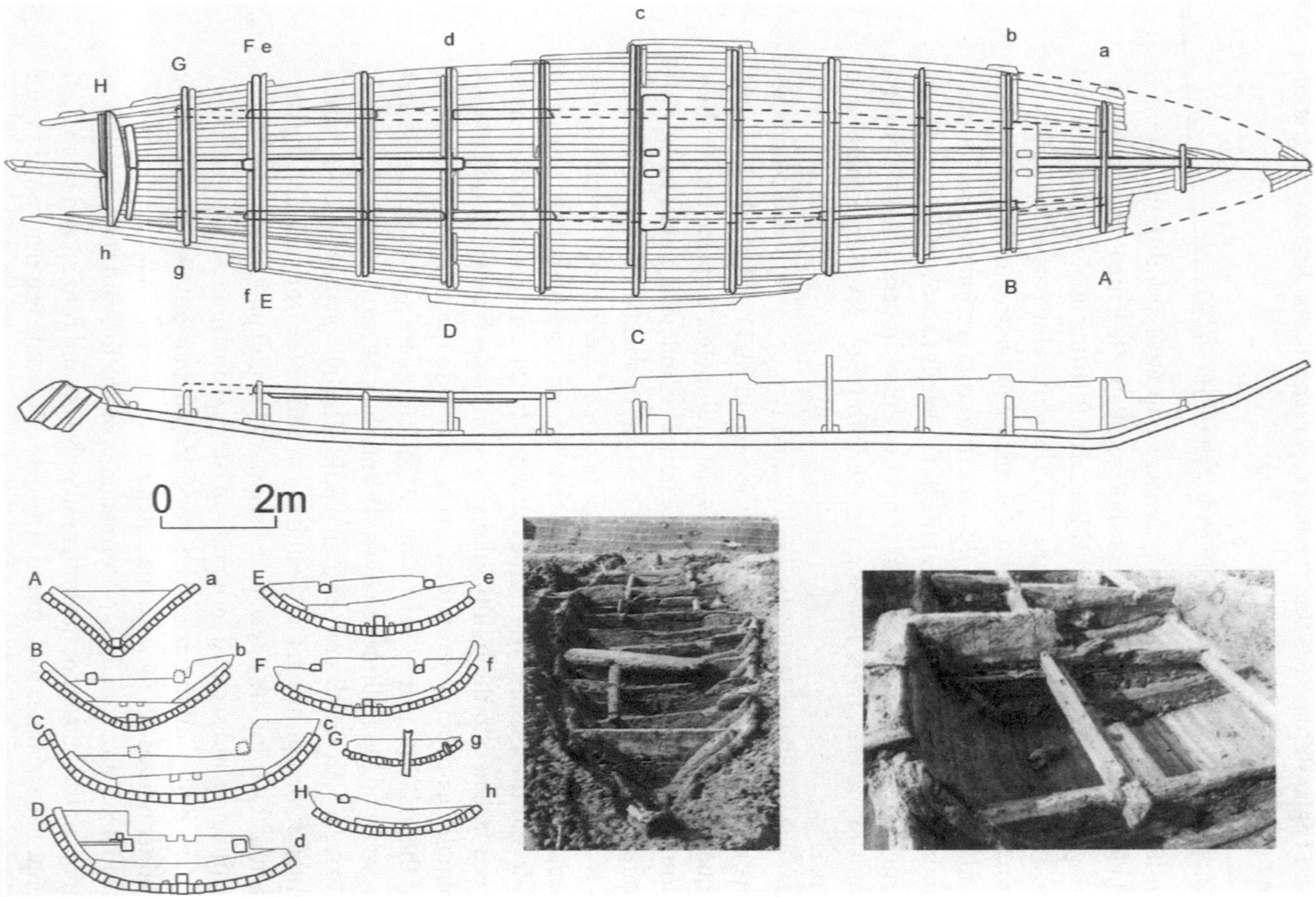

Figure 7.3. Xiangshan Ming dynasty shipwreck, plan and cross sections of the hull and two photographs showing the internal structure of the hull having bulkheads with longitudinal baulks. Letters indicate bulkheads and corresponding cross sections/views. (From Xi et al. 2004).

the early Ming dynasty. The nature of the ship's use has not been identified: it was either a local cargo ship used by private merchants or a battleship employed by the government. In most respects, the ship is regarded as similar to Penglai ship no. 1 (Xi et al. 2004: 199–200).

Penglai Ship No. 2 (蓬莱古船二号)

The remaining part of the hull of Penglai ship no. 2 measures 22 meters long and 5.2 meters wide. Although the stern is missing, the bottom of the hull is relatively well preserved, and the remains include a keel, some bulkheads, and eleven strakes on each side (figure 7.4). Details of the dimensions of the remaining planks are available in the archaeological report along with other detailed information about the hull (Cultural Relics and Archaeological Institute of Shandong Province et al. 2006: 25–34).

The cross section of the hull bottom shows some hollowing around the bow, and this changes to a rounded shape through the midsection. The keel, which shows slight hogging, comprises three members, with the forward keel measuring 4.78 meters in length and the main and aft keel measuring 16.22 meters. These are joined by hooked scarf joints with mortises and tenons. In the joints, there seems to be a hole for the placement of a mirror or coins, evidencing the practice of the shipwrights' beliefs. Iron straps and iron nails are used to fasten the forward and main keels. A hog piece is fitted on top of each joint, and large iron nails are driven from them down through the keel. The sides of the main keel are rabbeted, and thick planks (garboard planks) are fitted using mortise-and-tenon joints.

Although the ship originally had thirteen bulkheads, only six remain. There are limber holes from bulkheads no. 3 to no. 7, yet bulkheads nos. 2 and 8 do not have limber holes, because of the mast steps located at their forward faces. The lower and upper parts of each bulkhead plank are fashioned for a tongue and groove joint, and they are joined with mortises and tenons (loose tenons). Skewed iron nails are also driven into the seams of the bulkhead planks. Frames are used to fix the bulkheads and hull planking together, fastened by iron nails. The use of iron brackets to fasten the hull planking to the bulkheads is evidenced on bulkheads nos. 2, 3, and 7, although these brackets are mostly degraded and only corroded iron remains in the bulkhead plank surface recesses into which brackets were slotted.

The joint pattern of the hull planking shows the same method as in Penglai ship no. 1, with the use of large edge-driven nails and bent skewed

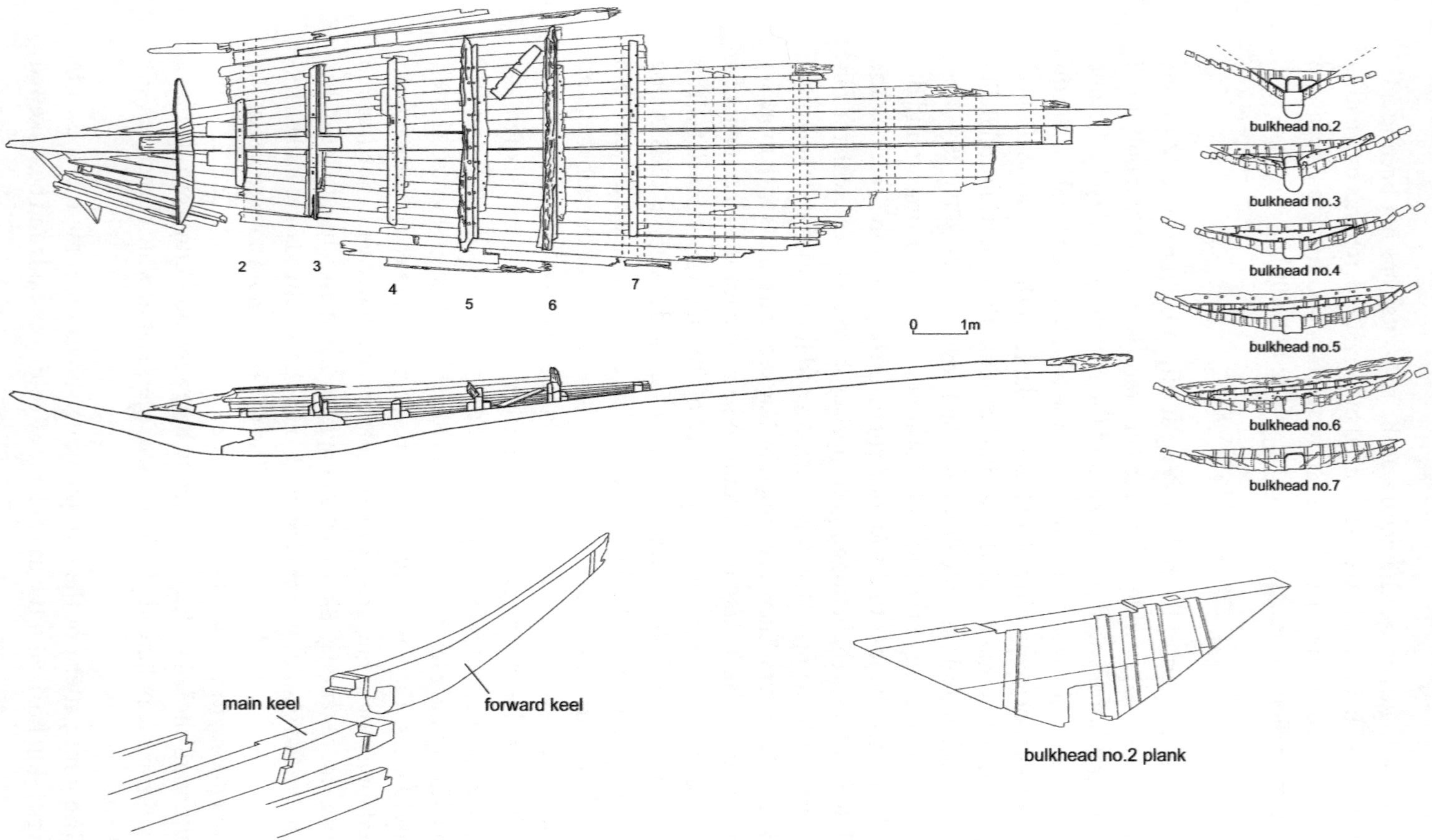

Figure 7.4. Penglai ship no. 2, plan of the hull. (From Cultural Relics and Archaeological Institute of Shandong Province et al. 2006)

nails. The mast steps are fixed to the hull planking by iron nails. The forward mast step measures 1.90 meters athwartships and has two square recesses measuring 160 × 225 millimeters. The mainmast step is missing. Penglai ships nos. 1 and 2 were evaluated as the same type of ships.

Penglai Ship No. 3 (蓬莱古船三号)

Despite the proximity of Penglai ship no. 3 to Penglai ship no. 2, the two ships are distinguishable in their structure. The remaining portion of the adjoining hull, identified as Penglai ship no. 3, measures 17.1 meters long and 6.2 meters wide; the remains include bottom planks, bulkheads, and hull planks, though the stern is missing (figure 7.5). Details of the dimensions of the remaining planks are available in the archaeological report along with other detailed information about the hull (Cultural Relics and Archaeological Institute of Shandong Province et al. 2006: 35–44). The structure of the hull bottom and planking shows remarkable similarity to those of indigenous Korean shipwrecks discovered in Korean waters (see chapter 2 and the final section of this chapter).

The hull has a flat bottom consisting of three strakes. The planks of the strakes are joined together by transverse wooden bars or dowels. Thirteen bars are used in the extant portion of the ship's bottom. Wooden nails seem to have been driven from the center bottom planks to the side planks in a few places. The structure of the bottom planks and their fastening methods are similar to those observed in excavated ships of Korean origin. However, the presence of an extra plank placed on top of the center bottom plank (and its fastening using edge-driven iron nails) has not been found on Goryeo period Korean ships.

Three strakes remain on the port side, and nine strakes remain on the starboard side. The first strakes are fitted into rabbets on the upper edge of the side bottom planks and appear to be fastened with mortises and tenons (or dowels). The result is a form of clinker-built hull planking with rabbeted seams. Skewed wooden nails are driven from the outer surface of the upper planks into the lower planks through the rabbeted seams. Butts of the hull planking are held together by lap joints. Also, iron nails are used for fastening the hull planking.

Bulkheads, which are not used in the Korean shipbuilding tradition, are the ship's main structures for transverse strength. Five bulkheads (regarded as the second to sixth bulkheads) remain, but more than eight may have originally been present. The bottom bulkhead planks have a recess

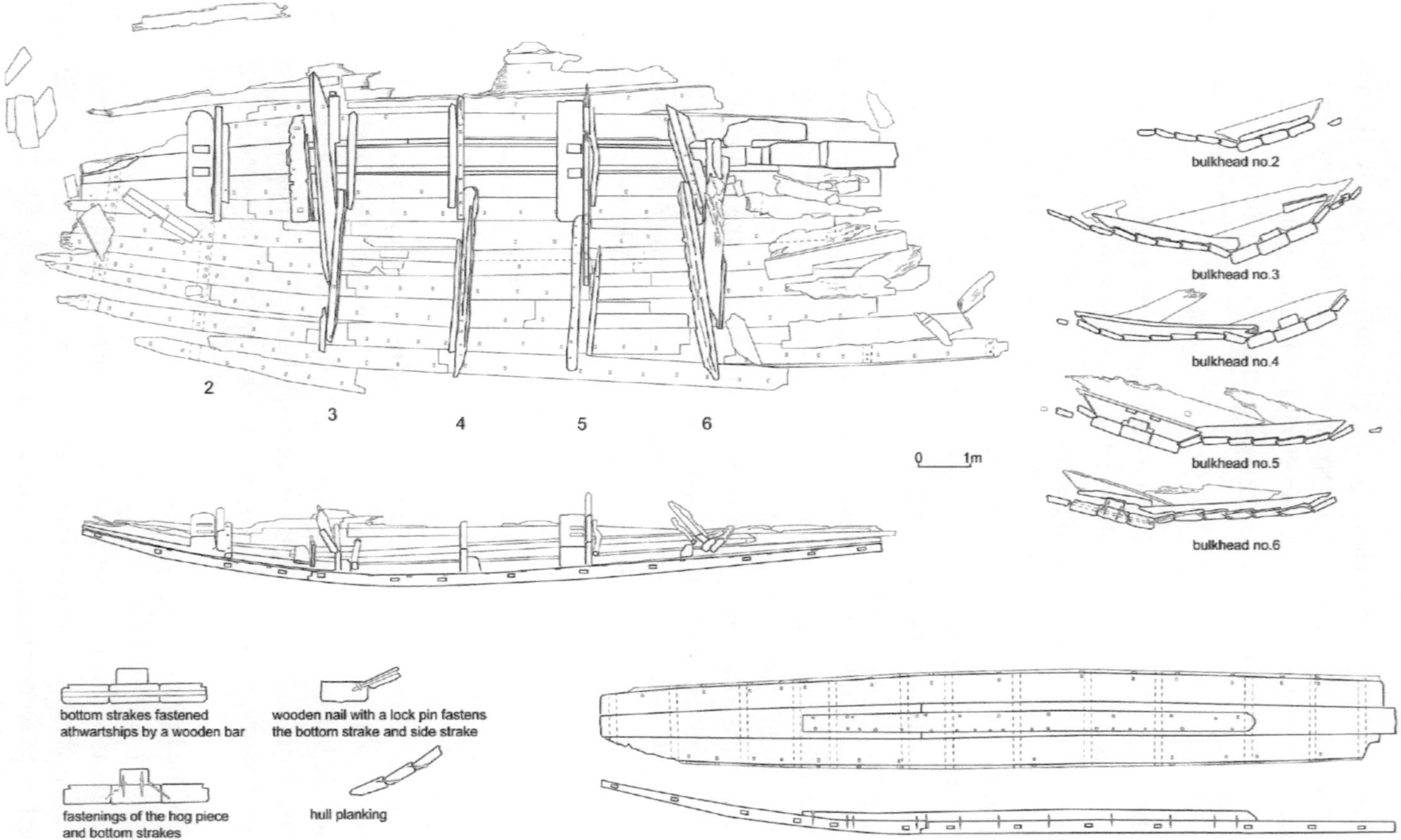

Figure 7.5. Penglai ship no. 3, plan of the hull. (From Cultural Relics and Archaeological Institute of Shandong Province et al. 2006)

in their bottom that fits to the extra plank on the center bottom, and they also create limbers. Bent skewed iron nails are used to fasten bulkhead planks as well as to fix the hull planking and bulkheads together. Frames are attached to some of the bulkheads, and bent skewed iron nails are used to fasten the hull planking to the frames.

Intensive use of iron nails is unusual in traditional Korean shipbuilding. The use of different types of wooden bars, mortises and tenons, and iron nails is emphasized in a comparative study on Penglai ships nos. 2 and 3 (Yuan 2006a: 500–501). The forward mast and mainmast steps, each with two recesses, are identified at the second and fifth bulkheads. The ship was discovered at almost the same elevation as Penglai ship no. 2, suggesting contemporary use of the two ships.

Penglai Ship No. 4 (蓬莱古船四号)

A few ship timbers were discovered about 12 meters away from the no. 2 and no. 3 ships. Only five timbers remain, yet they clearly show features identical to those of Penglai ship no. 3. According to an archaeological report (Cultural Relics and Archaeological Institute of Shandong Province et al. 2006: 44–46), three of four timbers found near each other are bottom planks forming a flat bottom. What remains of the center bottom plank measures 3.46 meters long, 200–440 millimeters wide, and 160–200 millimeters thick. It has two recesses that could receive tabernacle heels directly, instead of their being placed in a mast step.

Two bottom planks are positioned next to the center bottom plank, one on either side. The remaining part of each of these two planks measures 4.8 meters in length, 260–520 millimeters in width, and 100–220 millimeters in thickness. The upper edges of their outer parts appear to be recessed to place strakes. The three bottom planks are fixed by two wooden bars. One additional timber regarded as part of the hull planking was found nearby, and another large timber measuring 9.24 meters long (found away from the four timbers) is also regarded as a hull plank. Whether Penglai ship no. 4 had hybrid features like those of Penglai ship no. 3, in terms of bulkhead structure and the use of both iron and wooden fastenings, is unclear.

Characteristics of the Penglai Ships

Besides many structural features distinguishing the ship remains exhibiting elements of Chinese traditions (Penglai ships nos. 1 and 2) from the ship remains exhibiting elements of Korean traditions (Penglai ships nos. 3 and 4), the assemblage of woods used for each tradition also shows differences (see chapter 6). Penglai ships nos. 1 and 2 have more different types of wood used in the hull structure. In contrast, there is less diversity in the types of woods used for the main structure of Penglai ship no. 3; most parts of the hull are pine.

Study of East China Sea Ship Structure

The discovery of diverse types of ships at Penglai that show structural features originating from both the East China Sea and the Yellow Sea is indicative of long-term interactions in shipbuilding between the regions within East Asia by the Ming dynasty period. The hull having hybrid features is particularly significant. The existence of diverse technologies could have been a factor that facilitated the emergence of hull hybridization. A review of excavated ships dating as late as the Ming dynasty in China can lead to a clearer understanding of the diversity of coastal and seagoing ships. The ships excavated in East Asia have already been reviewed inclusively to identify structural characteristics (Green 1986; McGrail 2001; Van Tilburg 2007). The following section weaves together comparative insights about hull components used for subdividing the shipbuilding technologies within the sphere of the East China Sea tradition.

Hull Component Comparative Study

The condition of excavated ships, though variable, consists mainly of partial hull bottoms accompanied by other parts that have survived, such as bow, stern, and sides. All the identified bottom remains found in East Asia so far measure less than thirty meters in length, including the best-preserved hull remains of the Ming dynasty ships discussed above. The surviving parts of ship remains in the region, however, are sufficient for comparing their similar and idiosyncratic features.

The ships can be categorized based on the shape of the hull bottom with a focus on the cross section. The central cross section of historical East Asian ships is classified into one of the following generic categories:

Flat bottom (e.g., Tang dynasty riverine ships)
Floor with chine (e.g., Goryeo dynasty Korean shipwrecks)

Round bottom (e.g., Ningbo ship, Penglai ships nos. 1 and 2, Xiangshan ship)
V-bottom hollow deadrise (e.g., Quanzhou ship)
V-bottom sharp deadrise (e.g., Shinan shipwreck)

The differences in cross section are associated with structural distinctiveness, which is more clearly understood in the appearance of the seagoing ships that do not have a flat-bottomed structure. The following discussion highlights the hull features of such seagoing ships built in China to present an interpretive summary based on what has been observed through archaeological examination of hull components.

Keel

The development of the keel structure in the East Asian shipbuilding traditions is an important factor. The period when the keel was introduced into Chinese shipbuilding technology has not been determined, nor has the process of its incorporation into the ship structure. While we can assume that the technology for constructing ships with keels took time to be integrated into Chinese ship construction methods, the use of keels appears to have been adopted in central and southern China within a short time, perhaps before the tenth century. However, we cannot confirm this until the archaeological discovery in China of a seagoing ship of East Asian origin dating back to before the tenth century.

A keel is present in excavated ships dating to the Song and Yuan dynasties and engaged in maritime trading along the coast of central and southern China and across the East China Sea. Garboard planks were almost vertically joined to the keel, so that the extended hull planking forms a V-bottom hollow deadrise and a V-bottom sharp deadrise. Compared to the V-shaped bottom of the Quanzhou ship and the Shinan shipwreck, the deadrise angle of the Song dynasty's Ningbo ship is less steep. Its cross section forms a configuration similar to that of the Yuan to Ming dynasty Penglai ships nos. 1 and 2. Discussion regarding the similarity shows the paucity of evidence, as these ships are dated to later periods than the Ningbo ship, and they are from provinces farther north. Because of the destruction of the hull of the Ningbo ship, details such as its structural features and exact date can no longer be assessed.

The use of three members for the keel has been commonly identified on these ship remains (with a possible exception in the structure of the Ningbo ship). The length of the main keel in the five excavated ships ranges from 7 to 17 meters: the Quanzhou ship, 12.40 meters; the Shinan shipwreck, 11.27 meters; the Ningbo ship, 7.34 meters; Penglai ship no. 1, 17.06 meters; and Penglai ship no. 2, 16.22 meters. In previous studies, most of these ships have been estimated at slightly over 30 meters in length; for example, reconstructed hull lengths are 33.5 meters for the Shinan shipwreck and 31.1 meters for Penglai ship no. 2. When only the main keels are compared, however, the Penglai ships show a greater length, and the length-to-beam ratio of these two ships is also greater. These ships are said to have been used as patrol boats or battleships, and the sharp design of the hull appears to substantiate that use. However, whether the hull of the traders in the later periods had a wide beam as great as that of the Song dynasty traders may be worth examining in future study.

Bulkheads

Chinese ships became prominent in seafaring only after the ninth and tenth century. Excavated Tang dynasty ships illustrate the early use of bulkheads on riverine craft, and bulkheads continued to be used structurally as main transverse components when adopted in the construction of seagoing ships. The intended use of bulkheads to enhance transverse strength in the hulls of Song dynasty ships is, however, slightly different from the initial design of bulkheads to produce compartments in the hulls of Tang dynasty ships.

The bulkhead planks could have been sufficiently sturdy against crosswise distortion of the hull. The excavated ships show bulkheads that were probably strong enough to support the longitudinal and transverse strength of the hull, being fixed firmly to the hull through the use of brackets (iron brackets in the Quanzhou ship and wooden brackets in the Shinan shipwreck) and frames. The joint of the bulkhead planks appears to have been capable of resisting the pressure of seawater in a flooded hold. While this part of the purpose of the bulkhead structure is open to discussion, another feature to note is that the bulkheads of the Shinan shipwreck used vegetable fiber for caulking the plank seams. This watertightness would have secured the hold locally and protected cargo from getting wet. Leaking water was probably an unpreventable problem. The watertight bulkheads of the excavated ships were not fully watertight compartments; rather, their relative watertightness would have prevented the hold and cargo from becoming waterlogged.

Limber holes were needed in the bulkheads of the Song dynasty seagoing ships, while this feature seems to be absent in the excavated Tang dynasty rivercraft. The innovation of the bilge system is evident in the remains of Song dynasty ships: without limber holes in bulkheads, each bulkheaded segment would have needed a pump in it; if the limber holes were normally secured, then only one pump or man would have been needed to work at the lowest point in the hull. So far, there is no clear archaeological evidence of a system for pumping out water that collected in the bilge. In Osamu Oba's study of the records of Chinese traders that drifted onto the Japanese coast during the nineteenth century, there is a brief explanation about the management of these ships (Oba 2001). According to Oba, the original text of the record mentions a negotiation between Chinese merchants and sailors and Japanese authorities for repairing a hull, including a request to Japanese authorities to borrow a pump.

Analysis of the bulkheads in the Quanzhou and Shinan ships indicates that the bulkhead of Song dynasty ships should be interpreted as a component functioning to increase the transverse strength of the ship's structure, rather than to produce partitioned space or watertight compartments. For the bulkheads to have functioned effectively as transverse components against lateral pressures, half frames and brackets were used. The half frames did not function solely to increase transverse strength, and, similarly, the brackets were also used to secure the assemblage of bulkheads and hull planking. The complementary role of the brackets that secured the bulkhead planks suggests that they were most likely fixed into bulkheads that had already been assembled.

Seán McGrail (2001) has analyzed the arrangement of the brackets' installation to address the building sequence used for thirteenth- and fourteenth-century Chinese seagoing ships. In assessing the brackets used on Penglai ship no. 1, McGrail (2001: 372) noted that they are embedded within the thickness of the planking. Positioning the brackets this way suggests that the bulkheads had been positioned before the planking was fashioned, fitted, and fastened (McGrail 2001: 375). Penglai ship no. 2 shows a similar treatment of brackets: they are slotted between the strakes. The bent end is embedded, and the brackets are used in the same way between the keel and bulkheads (Cultural Relics and Archaeological Institute of Shandong Province et al. 2006).

Bracket installation in the Quanzhou ship and the Shinan shipwreck, however, is different from that of the Penglai ships; the brackets of the Quanzhou ship and the Shinan shipwreck pierce through the planking from the outside. The square holes or slits could be opened either before or after the fastening of the planking and bulkheads; that is, the system does not provide evidence of whether these ships were built by a bulkhead-first or a plank-first construction method (however, see chapter 4 for probable plank-first construction in the Quanzhou ship).

The distribution of bulkheads can prevent distortion of the hull cross section. In particular, racking (rolling) of the hull may cause deformation of the transverse section. A gale and waves cause tilting of the ship and generate substantial external force on one side of the hull. The force would affect the entire section, resulting in distortion of the hull. Appropriate distribution of transverse components, and tight fastening of these, can help reduce the effects of such forces. Bulkhead spacing is therefore significant in relation to producing sufficient transverse strength for the entire hull.

Analysis of the remains of the excavated ships, however, does not clarify how seriously the issue was addressed in their construction. The number of bulkheads in the hull differs substantially between the Quanzhou ship and the Shinan shipwreck: the former has twelve bulkheads; the latter has seven. The difference does not correlate with the proportions of the keel defining their distribution, because the total length of the main keel and the aft keel, where most bulkheads sit, differs by only about two meters between the two ships. As a result, keel proportion should not be regarded as an element affecting the difference in bulkhead spacing. It may be valid to say that the wide bulkhead spacing of the Shinan shipwreck is a distinctive feature. For example, the pattern of bulkhead spacing in Huaguang Reef shipwreck no. 1 appears to be as narrow as that of the Quanzhou ship. There are eleven bulkheads in the remaining part of the hull of the Huaguang ship, which measures approximately seventeen meters in length. The comparatively fewer bulkheads of the Shinan shipwreck may indicate that that hull had weaker transverse strength. The longitudinal and transverse structures are correlative in determining overall hull sturdiness.

This is important, because longitudinal and transverse components are interdependent structures and contribute synergistically to hull strength. Bulkheads appear to have contributed indirectly to longitudinal hull strength, for example, by supporting longitudinal baulks. Bulkhead planks from the Takashima underwater site have two square recesses that could receive baulks. A similar structure is seen in the bulkheads of the Ming dynasty Xiangshan ship, in which support baulks run longitudinally through the inside of the hull from the second to the twelfth bulkhead. Either the bottom or the second planks of each bulkhead have two square recesses horizontally at the same elevation to receive the baulks. Chinese researchers have concluded that the role of the baulks was to support lower decks in the hull (Xi et al. 2004). Because of the loss of the original deck structure, however, we cannot ascertain whether baulks related to decks, but the baulks could have contributed to longitudinal hull strength.

Planking

Improvement of planking techniques could have been an important innovation in Chinese shipbuilding, considering the strains that are placed on the hull of seagoing ships. The strain that causes longitudinal bending becomes more problematic in constructing larger wooden ships for long-distance voyages (Biles 1908). Under those circumstances, longitudinal strength of the hull becomes more important in constructing these ships. Assessment of longitudinal strength must take into consideration two stresses: bending moment and shearing forces. Longitudinal bending occurs when the ship is in the trough between two waves or on the crest of a wave, and the stress substantially impacts the keel. Distortion of the hull by shearing forces is another problem related to the longitudinal strength of the hull. To resist shearing forces, not only the keel but also the strength of the planking are important for longitudinal support.

Rabbeted seams feature in the planking of the Quanzhou ship and the Shinan shipwreck. They form carvel seams or clinker seams. The rabbeted clinker seams make the hull planking appear clinker-like. A combination of rabbeted carvel seams and rabbeted clinker seams is observed on the hull planking of the Quanzhou ship, while the planking of the Shinan shipwreck does not exhibit this combination. In the combination, the outer planks are adjusted to overlay both the carvel seams and the clinker seams of the inner planks, and this holds the assemblage of the inner and outer planks together as a single unit. The tightly assembled inner and outer planks form a plank shell, which provides longitudinal strength to the hull. The hull planking of the Quanzhou ship could have been more efficient against bending distortions that would have placed stress on each strake. The planking of the ships consists of multiple layers, in which the outer planks are thinner than the inner planks. The outer planks are perceived as sheathing planks to protect the inner planks from marine borers. The double planking of the Shinan shipwreck mainly functioned for this purpose.

These features are not similarly identified on the planking of the round-bottomed ships (for example, the Ningbo ship, the Xiangshan ship, and Penglai ships nos. 1 and 2). These round-bottomed ships do not use sheathing planks. The single-layer planking is composed of baulks and thick planks. The first strakes (like garboards) are fixed into the sides of the keel, not placed on top of the keel. The cross-section profile of Penglai ship no. 2 shows the fastening pattern of its planking. Besides skew nails that fasten the seams of the planking, spikes are driven from the seam of the upper plank through to the lower plank. Using such penetrating spikes would have been difficult in the hull planking of the V-bottomed ships using thinner planks. The discovery of the Song dynasty V-bottomed ships revealed the existence of the technique of using sheathing planks by extending multiple planking layers, which had previously been evident only in historical texts. This technique, however, may not have been a widespread innovation in the Chinese shipbuilding tradition. The planking observed on the round-bottomed ships, another common type of seagoing ship in the East China Sea, likewise does not occur in multiple planking layers. The multiple planking layers of the V-bottomed ships is understood as a technique developed endemically during the Song and Yuan dynasties and has been attributed to ships from the southern areas (perhaps south of Ningbo) that conducted voyages in the East China Sea and probably sailed to the South China Sea.

Additional Components

Thus, shipbuilding industries along the coasts of China bordering the East China Sea advanced the technology of constructing seagoing ships during the twelfth and thirteenth centuries. During these periods, maritime connections between the Chinese coasts were well established and involved different kinds of ships. Apart from the hull structure, consistency in using iron fastenings and standardization in the types of wood used for timbers characterize the East China Sea shipbuilding tradition in East Asia.

Other construction techniques and technologies will need to be elucidated to address the prominent presence of this tradition. For example, an adhesive substance known as *chunam* was used as putty to assemble the planks tightly together, as well as for caulking the plank seams and luting the surface of the outer planks. Development of the use of putty as a sheathing method has not been thematically studied, as its use has chiefly been discussed in the context of sealing iron nails and caulking plank seams—investigating its long-term use is beyond the scope of this chapter. To say that the use of putty was an essential technique developed on the Chinese coasts is, however, valid. While putty was consistently used even in the later East China Sea shipbuilding tradition, we need to examine whether its usage differed in various time periods coincident with historical transitions in shipbuilding technologies.

Ryukyu's Involvement in the Maritime Network

The historical importance of shipping in the Japanese archipelago stretching from the tip of southern Kyushu to Taiwan warrants mention. The island arc, known in Japan as the Nansei Islands but elsewhere called the Ryukyu (Okinawa) Islands, forms the southeastern perimeter of the East China Sea region under discussion in this book. Ryukyu's active maritime trading with China has been addressed in relation to a possible wreck site where ceramic sherds were found in the waters (Pearson 1969). More than fifty islands compose the Ryukyu chain, and the dates of the trade ceramics discovered there range from the tenth to the seventeenth century (Kamei 1993). The trading system of the Ryukyu Islands was formerly considered to have developed via Hakata in northern Kyushu, which was a major center of maritime trade with China and Korea. However, recent studies have indicated that early commodity exchange between China and the Ryukyu Islands probably has a bearing on the later period of seaborne trade, which allowed some chiefdoms on the islands to establish a direct trading relationship with coastal areas of China, such as Fujian, as early as the thirteenth century (Pearson 2009; Kinoshita 2014). Fujian celadon sherds dating back to the late twelfth or early thirteenth century, probably part of the cargo of a merchant ship, have been identified at the Kurakizaki underwater site at Uken on Amami Oshima Island (Uken-son Board of Education 1999). Recent studies of Ryukyu have assessed the route of the merchant ship, which could have voyaged directly from Fujian via the Ryukyu Islands, not northern Kyushu.

The rise of the Ryukyu kingdom (AD 1429–1879), based on the mainland of modern Okinawa, represents one of the many shifts in the maritime system in the East Asian realm. Trading with China officially required presentation of tribute to the Chinese court as a preliminary step. Ryukyu's diplomatic relationship with the Ming dynasty began in 1372 and involved more than 170 missions to the Chinese courts. The islands maintained independence by acting as intermediaries between the Ming and Japan, and they played a significant role in entrepôt trade (Wade 2007; Okamoto 2008b).

In examining the background of ships engaged in tribute trade, Hiromichi Okamoto has pointed out that they can be classified into three phases based on their origins and eras (Okamoto 2008a): ships commissioned by the Ming between the late fourteenth century and the early fifteenth century; ships built in Fukien on Ryukyu orders between the mid-fifteenth and early sixteenth century; and ships constructed in Ryukyu from the late sixteenth century onward. Differences in dimensions among the ships of the three periods are indicated by fragmentary information in historical texts on their cargo capacity; the seagoing ships built in Ryukyu could be larger than those of previous periods (Okamoto 2008a: 224–26).

Several iconographic sources and historical texts are available regarding nineteenth-century tribute ships that sailed from Okinawa to China (Okamoto 2008a; Tomiyama 2012). The iconographic resources provide realistic depictions of the Ryukyu tribute ships (figure 7.6), and the seaworthiness of the later Ryukyu tribute ships has been studied from the viewpoint of naval architecture (Yagi et al. 2008).

Japanese researchers have undertaken an archaeological search for tribute ships or sunken traders in Ryukyu waters. Scattered artifacts and items related to seaborne trade have been identified during a series of underwater investigations off the coast of the Ryukyu archipelago since 2004, led by the Archaeological Center of Okinawa (Katagiri 2007, 2009). Underwater explorations in 2010–13 located seven iron anchors and several intact local ceramics, the so-called Tsuboyaki (Ono et al. 2013). Determining the exact time of use of these ceramics is generally difficult, because they represent a local ceramic tradition that has prevailed through a long time period without modification in style.

Figure 7.6. Ryukyu archipelago ships: (*top left*) *shinko sen*, a tribute ship; (*top right*) *kai sen*, a ship voyaging to Kagoshima; and (*left*) *maran sen*, a local Ryukyu ship. (Courtesy of TNM Image Archives)

Iron anchors were found, but a typological approach to iron anchors has not yet been developed in detail in Japan. The anchors have four arms and range in size from 1.2 meters to 2.0 meters, and their association with the local ceramics suggests that they were from either Japanese traders or Ryukyu ships (Ono et al. 2013). The possibility that they might have originated from Chinese ships should be pursued as well, because envoy ships from China that visited Okinawa were clearly equipped with iron anchors, according to an iconographic source (Needham 1971).

Ship remains from Japan's Ryukyu Islands and from Chinese coastal waters can help cast light on shifts in the regional network of shipbuilding and maritime trade. Increased building of seagoing ships around the East China Sea during the Song dynasty was part of an effort to boost trade, and loss of northern territory moved the effort southward. The Mongolian rulers of the Yuan dynasty then brought dramatic changes, suspending the tribute system and provoking a surge in private trade—but piracy also surged as Yuan power declined. The subsequent Ming dynasty had to assert its naval power to expand its jurisdiction and restore imperial hegemony at sea. At the same time, a new type of maritime networks emerged, as represented by the rise of the Ryukyu kingdom.

8

East Asia's Link to the South China Sea and Gulf Traders

The South China Sea, which encompasses southern China, is rimmed by a series of Southeast Asian nations (map 8.1). Sea routes stretching from China via the South China Sea to the Indian Ocean have a long history that has undergone much scholarly review (Yule 1882). The historical sea corridors connecting East and West have been termed the Maritime Ceramic Route, similar to the Silk Road on land (Mikami 1969). Historical ports and trading points for ships laden with ceramics have been explored, and intra- and transregional trading have been reviewed more recently with shipwreck data (Heng 2009; Miksic and Yian 2013).

The structures of several shipwrecks found in the South China Sea and some gulfs in Southeast Asia are remarkably similar to the structure of seagoing ships excavated in the East China Sea. This fact is tightly linked to my theme of technological hybridization, the processes of which I consider can be understood more precisely by linking the formation of shipbuilding traditions in these regions with a chronological perspective. Highlighting the issue leads to a hypothesis about the possible influence of construction methods that originated in East Asia on shipbuilding in the South China Sea. Early seafaring and endemic ships in the East China Sea and South China Sea can be described based on the historical records of Chinese envoys and pilgrims (see the following section). Notably, these records illustrate the early presence in the South China Sea of Southeast Asian indigenous ships and ships from the Indian Ocean, well ahead of the appearance of Chinese ships.

While the historical sources lack detailed information on the physical features of ships and the techniques used for their construction, the recent increase in archaeological excavation of ship remains in the South China Sea provides important clues to understanding details of early shipbuilding technologies. Focusing on findings representative of Southeast Asia is

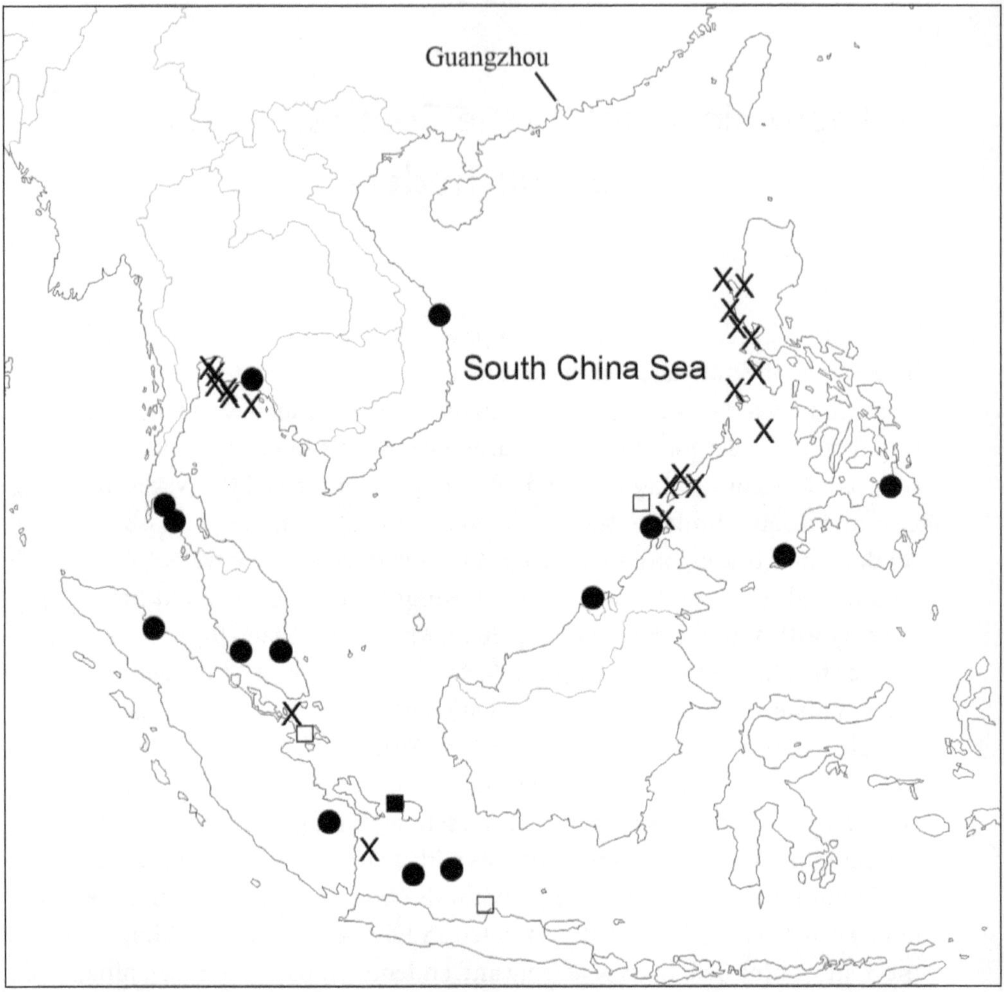

Map 8.1. Map of the South China Sea region showing the location of assorted ship remains and wreck sites. See appendixes 2 and 3. (Produced by author, after CraftMAP)

● indigenous ships with lashed-lug technique (third–fourteenth centuries)
■ Indian Ocean seagoing ship (ninth century?)
□ East China Sea traders
X Gulf/South China Sea traders from Thailand, the Philippines, and Indonesia (fourteenth century and later).

important for understanding the developing intersection of shipbuilding traditions in southern China and foreign technologies, including the early Southeast Asian and Indian Ocean shipbuilding traditions.

East China Sea shipbuilding traditions (addressed in previous chapters), came to be influential in the South China Sea region after the tenth century. Although how ships from the East China Sea and ships endemic to Southeast Asia coexisted is poorly understood, shipbuilding innovations that originated in East Asia became an integral part of shipbuilding in the Southeast Asian regions of the South China Sea. The aspect I wish to emphasize is that in parallel with the cultural and social independence of the South China Sea region, the shipbuilding tradition on the southern coast of China facing the South China Sea was in a different sphere from the traditions operating along the other parts of the Chinese coast. The powerful independence of the south was supported by maritime trade linked to Southeast Asia and even the Indian Ocean, rather than to northern China. This maritime trade in the South China Sea, known as the Nanhai trade, helped local officials accumulate considerable wealth—but their market could exist only through the demand of the central rulers. Through most of the first millennium, local societies had not matured sufficiently to demand valuable consumer materials. Under these circumstances, Chinese merchants would probably not have seen much to gain by being involved in maritime trade on their own, possessing their own trading ships and organizing their own sailors. Rather, their role involved transporting goods from the south to the north in local coastal trading. The idiosyncrasy of the south in its other maritime activities persisted throughout much of the first millennium; the far south was not significantly sinicized (brought under the influence of Han Chinese rulers). However, socioeconomic expansion and changes of polities on the Chinese mainland during the Song and Yuan periods facilitated unity along the coasts of the East China Sea and South China Sea. Consequently, the emergence of a distinctive new shipbuilding tradition could virtually coincide on the mainland coast and in insular Southeast Asia. The exact timing of the period of innovation is disputable but likely dates back to as early as the fourteenth century.

Archaeologically, several shipwrecks dating to the fifteenth century illustrate and confirm the technological innovation represented by a hybridized hull. The Gulf of Thailand was the location for the main suppliers of these ships, partially because of the long-term Siamese relationship with the Chinese mainland (Viraphol 1977). Known as gulf traders, in later periods they became prominent even in transregional maritime trade reaching East Asia.

Early Seafarers in the South China Sea

The South China Sea has historically been a zone of intermingling, where different groups of people have been involved in complex maritime activities and shipping. The people contributing to the development of ship structure and construction are not limited to the carpenters who worked in shipbuilding industries. We also need to consider those who demanded oceangoing ships as a form of transportation for commodities and people. Historical texts may not provide evidence about these people, who appear only in fragmentary information associated with records of journeys, immigrant processes, military activities, and trading.

For the first half of the first millennium, we have scarcely any descriptions of movements and settlements relying on ship transportation. It has been noted that around AD 132 the coastal area stretching from southern China to northern Vietnam was an important zone of early contact between northern Chinese rulers and early Southeast Asian rulers (Chang 1934). The claim has been made, on the basis of limited evidence, that "Chinese ships were reported at Penang on the west coast of the Malay Peninsula by 350 A.D. and at Ceylon by the end of the 4th century" (Guy 1980: 11). Settlement of sinicized Chinese around the South China Sea before the tenth century is barely evident in the historical texts (Wang 1958: 93). The only reliable source that mentions Chinese merchant settlement before the Song dynasty is a travel record of the Tang dynasty Buddhist monk Iching (義浄 [Wang 1958: 93]). His account is in the *Da Tang Xiyu Qiu Fa Gao Seng Zhuan* (or *Ta Thang Chiu Fa Kao Seng Chuan*, Account of Buddhism sent from the South Seas and Buddhist monks' pilgrimage of the Tang dynasty, 大唐西域求法高僧傳).

Apart from the monks' pilgrimage records, there are few other historical texts, and those are formal records of seaborne trade as part of the tribute system of Chinese dynasties. Before the ninth century, the tribute trading system served in diplomacy between central China and neighboring countries and facilitated cultural exchange with states in the South China Sea. Safe water transportation between China and these states must have been an important part of the tribute and cultural exchange system.

A query arises about who were the voyagers employed for state traders that sailed between the two regions. The activities of Chinese merchants in the maritime trades in the South China Sea are said to be traced up to the mid-tenth century in historical texts (Enomoto 2008). But who was engaged in the Nanhai trade before that time? Gungwu Wang, for example, has insisted that several groups of people were involved in maritime activities from the first century BC to the founding of the Song dynasty in the tenth century (Wang 1958: 113). The stages of activity can be divided into the following phases:

1. The first five centuries up to the end of the Jin dynasty (AD 265–420), when Indian merchants and the Yueh people were the predominant providers of southern goods to northern rulers.
2. The two centuries of the Southern Dynasties (AD 439–589), when there was a need for "holy things" such as jewels, perfumes, and religious items, which were carried by Bosi (Po-sse) and Kunlun traders and sailors from the Southeast Asian coasts and archipelagos through contact with Java-Sumatra and Ceylon.
3. The period from the Tang dynasty to the rise of the Song dynasty, when massive quantities of medicines and spices were demanded. The commercial power of Srivijaya (as it was known in the seventh to thirteenth centuries [now called Sumatra]) and the embarkation of Persian and Arab merchants and sailors occurred in the early stage of this period. This finally stimulated the advance of people from the mainland of China into maritime activities in the South China Sea.

Maritime Power in Southern China

Indigenous seafarers in southern China successfully conducted long-distance voyages into the deep waters of the South China Sea. However, because of a lack of sources, the details about the involvement of these seafaring groups are highly speculative. Rulers in the northern part of China are known to have recognized the existence of such groups early on and to have seen the potential in establishing supremacy in the south. Successive central Chinese rulers tried to intervene in the southern part of the continent not only for territorial expansion but also for direct access to the South China Sea.

This was initially achieved by Zhao Zheng, the ruler of Qin; during his reign (246–210 BC), he founded three commanderies in southern China in what are now the provinces of Guangdong and Guangxi (Kuwata 1954). However, Zhao Tuo, the military commander of the Nanhai Commandery, began to fight for independence and successfully established a separate kingdom after defeating the kingdom of Nanyue (Nanyueh, Vietnam) in 204 BC. Occupying most of southern China as well as northern Vietnam, the new kingdom became extremely prominent (Kuwata 1954: 5–6), partly because of the economic interests nurtured by maritime trading linked to the South China Sea. The capital of the kingdom was located in what is now Guangzhou in Guangdong, where an ancient shipyard has been discovered near the tomb of the second king of Nanyue.

Excavation in the 1970s revealed wooden structures, regarded as the remnants of slips, with a length of twenty-nine meters. Radiocarbon dating conducted on the wooden remains in 1976 dated the first slip discovered to 2190±90 BP (240±90 BC), and analysis of the strata and sedimentation identified the period of active use of the shipyard as from the Qin occupation to the reign of Nanyue (203–111 BC [Guangzhou Shi Wenhuaju 1999]). No ship remains were found at the site. The archaeological report includes a discussion of possible ships built in the shipyard, based on wooden shipbuilding traditions in modern Guangzhou. There has been almost no consideration, however, of the possibility that the traditions of seafarers and ship carpenters in the area during the Nanyue period might show similarity to those of Southeast Asia (Austronesia) in terms of skills and technology.

The reign of Nanyue's rulers should not be underestimated, but the imperial courts and rulers based in northern China intervened in an internal Nanyue feud. In a record of the expedition of Emperor Wu of the Western Han dynasty to the south is a depiction of a ship called *lou-chuan* (castle ship, 楼船), which was used in battle against Nanyue (Kuwata 1954: 5–6; Needham 1971: 424–26; Xi et al. 2004: 51–53). Southern interventions by two Han dynasties allowed them access farther west. According to Wang (1958: 21), "It may be claimed that there was trade between Han China and the countries of the Nanhai in the years following the fall of Nanyueh at the beginning of the first century B.C. and that this trade had been gradually extended to include the areas to the west India [*sic*]." The southern port cities were the frontier of the Chinese empires, but they were also the gateway to the South China Sea. The Han dynasty could send imperial envoys to the southern seas with assistance from skilled Nanyue sailors (Guy 1980: 11), and later the dynasty flourished with the arrival of foreign ships.

Historical Text Fragments on Early Voyagers and Their Ships

Groups such as the Funan and Kunlun were recognized as authentic seafarers who lived on the coasts of mainland Southeast Asia and voyaged into the waters of the South China Sea. The South China Sea was a corridor between East Asia and the Indian Ocean world, and those seafarers played significant roles in transit between the two regions (Yamamoto 1980; Hall 1985). During the third and fourth centuries, the early international trade route from China to the Malay Peninsula ran along the Gulf of Thailand (Siam) and the Funan coast. An ancient Khmer kingdom called Funan (AD 68–550) occupied the area around the Mekong delta and rose to prominence in the maritime trade between China and India; the significance of Oc Eo as an ancient Funan transregional port is mentioned in several sources (Bellwood 1985; Higham 1989; Glover and Bellwood 2004; Manguin et al. 2011). Wu (AD 229–80) dispatched to Funan two of his commanders, Zhu Ying and Kang Tai (朱應, 康泰; Watanabe 1985), according to the historical Chinese documents *Funan Yiwu Zhi* (Record of rarities of Funan, 扶南異物志) and *Wu Shi Waiguo Zhuan* (Foreign countries in the period of the kingdom of Wu, 吳時外國傳). The record of their third-century journey indicates that they traveled through many countries in Southeast Asia; the original texts are now lost, but fragments are present in texts written in later periods (Watanabe 1985: 7). Kang Tai's report, *Wu Shi Waiguo Zhuan*, mentions the configuration of Funan ships:

> In Funan, timbers were cut down to construct ships, measuring twelve *hsin* [about twenty-three meters] long and six *shaku* [about three meters] broad, with iron sheets, and their bows and sterns were shaped like fishes. The large ones carried a hundred men, each man carrying a long or short oar, or a boat-pole, from bow to stern, fifty men sitting on a side, or forty-two men, depending on the size of the ships, using the long oars when standing, using the short oars when sitting down, in sailing on the shallow waters using the long boat-pole. (*Thaiphing Yulan*, vol. 769: *Li Fang* [author's translation])

扶南国伐木爲舡，長者十二尋，廣肘六尺，頭尾似魚，皆以鐵鑷露裝，大者載百人，人有長短橈及篙各一，從頭至尾，面有百五十人，作或四十二人，随舡大小，立則用長撓，坐則用矩橈，水淺及用篙.（太平御覧，巻769, 李昉)

This description partly illuminates the configuration of the long boat commonly used in Southeast Asian waters. Although it does not mention sails, there is an indication of the use of seasonal winds in Kang Tai's record. Another third-century historical passage, *Nan Chou Iwu Chi* (Record of rarities of southern regions, 南州異物志), indicates that "foreigners," meaning people outside the Chinese cultural sphere, used a sail on their ships. These ships, called *po*, measured sixty meters long and about six meters in breadth (although the dimensions appearing in the text could be exaggerated) and carried six hundred to seven hundred men. The text emphasizes that the sails on the foreign ships allowed them to sail under strong gales and violent waves with sails made of woven fronds. Multiple sails rigged from bow to stern were maneuverable to respond to wind direction.

The rise of Funan likely influenced the development of its local shipbuilding technology (Hall 1985). But although for a time Funan occupied an important position in the maritime network of the South China Sea, in the fourth century the political situation in Funan became unstable (Hall 1985: 68–77). The links between Han China and Funan and from Funan farther to the west were still important, but other sea routes between the two regions probably developed. The disorder in Funan resulted in a shifting of the sailing routes farther south, where around the fifth century other countries of the South China Sea, such as Java-Sumatra, gained power as entrepôts between China and India (Wang 1958). As this occurred, "the transhipment across the Malay Peninsula was abandoned in preference for sea travel through the Straits of Malacca to the western edge of the Java Sea and then north to China" (Guy 1980: 12).

The development of knowledge regarding the monsoons and sea currents of the South China Sea in early seafaring remained precarious, however, and there was still substantial risk on long-distance voyages. A pilgrimage in 411 by Fa Hsien (or Faxian) (法顯) to India and Ceylon is recorded in the historical text *Fu Kuo Chi* (Record of the Buddhist kingdoms, 佛國記) or *Fa Hsien Zhuan* (Biography of Fa Hsien, 法顯傳 [Wang 1958: 42–43]). On his return journey from Ceylon to China in AD 413–14, he appears to have taken passage on a Southeast Asian merchant vessel (Hall 1985). The voyage wound up in distress, but the ship finally reached a location identified as "Yeh-p'o-t'i"—possibly Java (耶婆堤) or the west coast of Borneo, according to Kenneth Hall (1985: 40).

> After receiving the Sanskrit verses (in Ceylon), I got on board a large ship carrying about two hundred people. . . . [After] being caught in a gale throughout thirteen days and nights, we managed to get to the shore of an island. At low tide, we repaired the leaking hull and sailed back to the waters. There were many pirates around this area. None of us would have survived if we had encountered them. The ocean was so vast that by determining the direction of east and west only from the position of sun and stars, we could barely sail. . . . Thus, after ninety days we finally reached a country. It was called Yeh-p'o-t'i. (*Fa Hsien Zhuan* [author's translation])
> 得此梵本已即載商人大舶上可有二百餘人．．．．如是大風晝夜十三日到一島邊．潮退之後見船漏處即補塞之．於是復前．海中多有抄賊．遇輒無全．大海彌漫無邊不識．東西唯望日月星宿而進．．．．如是九十許日．乃到一國．名耶婆提.(法顯傳)

The historical text shows, without mentioning specific navigational tools, that the sailors on Fa Hsien's voyage had no landmarks to follow but instead steered by the position of the stars and sun. The ship used to cross the Indian Ocean appears to have been a large Indian ship (Wang 1958: 43). Although the details of this ship are not specified, the role of such ships in long-distance voyages taking a direct route from the Indian Ocean to the South China Sea is implied by the context. After a journey of ninety days to get to Yeh-p'o-t'i, Fa Hsien had to spend five months there. When departing Yeh-p'o-t'i, he was transferred to another ship to reach Guangzhou. This ship likewise was said to be capable of carrying more than two hundred people. A group of Brahmans seem to have embarked together with him, but no details are provided that would clarify whether the second ship on which Fa Hsien took passage was Indian. This ship also lost its way, and the journey took seventy days, twenty days longer than the fifty days that were normally required for crossing northward to the South China Sea. Navigation techniques of the time consisted of primitive methods, not yet quantitative and mathematical as in later periods (Needham 1971: 562–63). Fa Hsien seems not to have seen other Chinese people during his stay in Ceylon. Apart from the travels of some officially sanctioned Chinese monks, as exemplified in Fa Hsien's journey, around the fifth century it was still unlikely for the Chinese to be actively involved in voyaging farther west of the Malay Peninsula and Indian Ocean.

Another Chinese Buddhist monk, Shiyun Wujie (釋曇無竭), whose name appeared in the *Gao Seng Zhuang* (Biographies of eminent monks, 高僧傳), traveled to India by land in AD 420 and retraced Fa Hsien's return route. On the way back to China, he may have embarked on one of the ships that departed from southern India via the South China Sea and reached Guangzhou (Watanabe 1985: 9). Awareness of seasonal monsoons and tidal currents is apparent in the voyages of these two Chinese monks; seafarers also could rely on the use of sails, which had developed by that time. However, from the two monks' descriptions of their voyages, it is difficult to glean information about the development of navigation techniques and hull structure.

An annotated text regarding a Buddhist scripture, known as *Huilin Yinyi* (Literary, phonetic, and semantic dictionary of Huilin, 慧琳音義), written between 783 and 807, offers a glancing description of the hull of a Southeast Asian vessel called a *kunlun* ship. The ship in the text appears to have been a locally built ship, not having originated in India. The name *kunlun* was used in Tang and Song dynasty records for ships collectively calling around mainland Southeast Asia from farther west. *Kunlun* is a general term applied to ships from the Malay Peninsula or, more broadly, the Indochinese peninsula. Water transportation routes developed over centuries along the coasts of the peninsula and connecting into the delta systems (Stargardt 1984). Beyond the assumption that the ships must initially have been built and used by the Kunlun people, we have no clear definition of these ships, which could have included various types and sizes of vessels operating along the coasts of the peninsula. The description in the above-mentioned text provides an impression of a ship of substantial size, but the more notable information is about the fastening of the hull planking, which had probably been constructed using some sort of wooden fastenings. Vegetable fiber appears to have been used to caulk the seams of the planks. The text does not clearly address whether vegetable fiber was also extensively used for other parts of the hull planking. The *kunlun* ship's planking struck the person who recorded it as unusual and as showing unique techniques. This suggests that the use of iron nails in shipbuilding was more typical in China wherever the person came from, likely farther north. In contrast, in some southern regions of China, such as the area of Guangzhou, during the early Song dynasty, hulls were still being constructed by sewn or lashed techniques using rattan.

> Large ships in the sea were called *bo*, and according to Guang Ya, all these ships were for seagoing purposes. The size of the ship was as large as fourteen meters, managing to carry over a thousand people excluding the cargo. It was also called *kunlun bo*, because Kunlun people were employed as sailors to sail it. Rigs were made from the bark of palm trees, and the woven fiber of kudzu was used for caulking, so there was no leaking of cold water. Iron nails or clamps were not used, because of the risk of fire from friction of the iron; instead, wooden fastenings fastened the hull planking. Passengers were anxious about using such thin planks for the hull. The hull would look as long as a mile, having the three members. Propulsion fully relied on sailing by wind, rather than manpower. (*Huilin Yinyi*, vol. 61 [author's translation])
> 海中大船曰舶廣雅舶海舟也人水六十尺使運載千餘人除貨物亦曰崑崙舶運動此船多骨論為水匠用椰子皮為索連縛葛覽糖灌塞令水不入不用釘鍱恐鐵熱火生纍木枋而作之板薄恐破長數里前後三節張帆使風亦非人力能動也. (慧琳音義 卷六十一)

Southern China may be regarded as a culturally independent region, the natural position of which placed it under the influence of both China and Southeast Asia before the tenth century. In completing successful voyages crossing the South China Sea, the Chinese monks aimed at getting back safely to ports in Guangzhou, China, which was the destination of the South China Sea trade on the Chinese continent around that time. The city, the most southerly gateway to China, was a hub port linked to Southeast Asia and places farther west. There were local voyagers skilled in shipping in the region, although they may not have been regarded as Han Chinese people. Yueh people, for example, could have been representative of those who had remained unsinicized as voyagers. Although details about them are not definitive, Yueh people had occupied the Guangdong and Fujian areas until the Tang dynasty. Indian people started to appear in the southern ports of China during the same that time.

Guangzhou could have been the most important international port in southern China by the tenth century, which resulted in the establishment of the so-called maritime supervisorates, or Shibosi, for the administration of merchants and ships from overseas (Kuwabara 1989). Terms for foreign ships that appeared in several Chinese historical texts included *ban bo* (蠻舶), *bolumen bo* (波羅門舶), *bosi bo* (波斯舶), *fanbo* (番舶), *hai bo* (海舶), *kunlun bo* (崑崙舶), *kunluncheng bo* (崑崙乘舶), *nanfanhai bo* (南蕃海舶), *nanhai bo* (南海舶), *shiziguo bo* (師子國舶), *waiguo bo* (外國舶), *xinanyi bo* (西南夷舶), and *xiyu bo* (西域舶 (Kuwabara 1989). Some of these names are simply general terms for foreign ships. Ships associated with specific groups of people or countries are *fanbo*, possibly outriggers from Southeast Asian archipelagos; *bosi*, possibly from Persia; *kunlun* (possibly Khmer ships from mainland Southeast Asia); *bolumen*, possibly from India; and *shiziguo*, from Ceylon, although further details of these ships are unclear.

Archaeological Perspectives on Early South China Sea Boatbuilding Techniques

Long-distance sea routes lying between the South China Sea and the Indian Ocean were maintained by the involvement of a range of seafarers, some of whom were engaged in transregional voyages. No hull remains related to early voyages to the South China Sea have yet been documented in India's coastal waters. In contrast, the existence of various types of indigenous boats is indicated by the discovery of numerous stone anchors and a wealth of underwater wreck sites in Indian waters (Tripati et al. 2001; Tripati et al. 2004). The configuration of early Indian sailing ships around the third century is inscribed on terracotta seals found at Chandraketugarh in West Bengal, India (Chakravarti 2002). A ship depicted among the famous cave paintings in India's Ajanta Caves, dating to the seventh century, provides further details of an early Indian sailing ship (Hourani 1963).

Indian seafarers were not the only distant travelers; ships from many different places—including Sri Lanka, Persia, and the Arabian region—were engaged in transregional seafaring. A major destination for these vessels and the people on board was Guangzhou in southern China. The ninth-century Belitung shipwreck discovered in Indonesia is a recent addition revealing early Arab contact with China. It is an important resource illustrating the direct voyage of a ship from the Indian Ocean to the South China Sea. The site is in a poor condition of preservation, and the site assessment revealed a scatter of broken ceramics, which may have resulted from commercial salvage at the site (Sudaryadi 2011). People consider that the ship originated from the western Indian Ocean. Use of African timber and similarity of construction techniques and design to modern Omani traditional boatbuilding are clues for addressing the origin of this ship (Tom Vosmer, pers. comm., 2010). The ceramics recovered from the shipwreck illustrate the active involvement of non-Chinese ships in early trading of Chinese commodities in the South China Sea (Guy 2006). Persian and Arab merchants were among the foreign merchants who appeared on the southern coast of China, where Guangzhou was the most renowned international trading port during the Tang dynasty, though during the Song and Yuan periods the bulk of maritime trading shifted to Quanzhou in Fujian and Yangzhou in Jiangnan, on China's central coast. These areas facing the East China Sea thus came to be a melting pot of technologies, where Chinese shipbuilders encountered the shipbuilding traditions that had advanced in the South China Sea and the Indian Ocean. The innovations resulting from the encounter led in turn to the advent of the East China Sea shipbuilding tradition in the South China Sea. This facilitated the advancement of the South China Sea shipbuilding tradition to the next stage of development.

Southeast Asian Lashed-Lug Frame Construction

Long-term use of long dugouts is widely observed on the mainland of Southeast Asia, as represented in the motif of bronze Dong Son drums and in the reliefs of Khmer monuments in the Indochinese peninsula. The primitive dugouts exhibit innovation partly in the utilization of dowels, or treenails, for adding planks. The enduring tradition of planked-up dugouts with dowels can be traced up to the twentieth century in many parts of Southeast Asia (Burningham 1989). Besides the planking methods using the wooden fastenings, a stitched technique with the use of organic fiber has been reviewed as a representative technique in the region (McGrail 2001; McCarthy 2005). The two techniques were combinable in early shipbuilding. The fact that a group in Taiwan speaking an Austronesian language was still using them into the twentieth century has been well documented, as a boatbuilding tradition similar to that of Southeast Asia (Kano and Segawa 1956; Deguchi 1995). This is noteworthy in that the early dissemination of shipbuilding and interaction of different traditions occurred widely within the Asia-Pacific regions, contributing to the cultural development of the Austronesia-speaking people (Bellwood 1985; Lape et al. 2007).

At some time before the tenth century, shipbuilding in the South China Sea reached maturity in the construction of seaworthy ships. Such ships had substantial loading capacity for human and cargo transportation, compared to planked-up dugout canoes. This advancement is associated especially with an innovation in the transverse structure: expansion of the beam and reinforcement by frames. The frames are fixed into the hull by a lashed-lug technique (Horridge 1982). The technique allows the frames to be tightened by fibers onto protrusions chamfered on the inner surface of the hull planks. The planking associated with the lashed-lug technique is principally an edge-joint method, initially used with a combination of stitching and wooden dowels and later with wooden fastenings only. There are at least two representative types of lug, based on the shape of the protrusion: a round lug and a rectangular chamfered protrusion. Lugs of the former type are seen in early ship timbers found in Pontian on the Malay Peninsula, while excavated boats in the Philippines known as Butuan (or *balangay*) boats represent the second type (Manguin 1993). Of the five Butuan boats found in 1986, three have been fully excavated. While the dates of the boats based on radiocarbon dating analysis are disputable, they have been fully assessed by several researchers in terms of structural details (Ronquillo 1987; Green et al. 1992; Clark et al. 1993; Green et al. 1995).

There has been an increase in the discovery of ship remains showing the lashed-lug technique across the South China Sea since 2000 (see map 8.1). Indonesia is the focal area for the discovery of ancient Southeast Asian ship remains, although most of them consist of disarticulated timbers (Manguin 1993). A remarkably well-preserved ship at the site of Punjulharjo, near Rembang in Java Tengah (central Java), has been reported (Manguin 2009; Priyanto 2011). The hull remains of the Punjulharjo ship measure 15.6 meters in length and 4.5 meters in width amidships (figure 8.1). The remaining part of the hull consists of a well-preserved bottom structure, including a keel, five strakes on the port side, and four strakes on the starboard side. Parts of the bow and stern structure also remain. A rudder post was discovered next to the hull.

The hull construction techniques evident on the Punjulharjo ship show that shipbuilders in the Southeast Asian boatbuilding tradition built

Figure 8.1. The Punjulharjo ship discovered in Java. (From Manguin 2009; courtesy of Ministry of Maritime Affairs and Fisheries)

seaworthy ships. The identified planking method involves the fastening of hull planks by both dowels and disconnected stitches. After the strakes were joined, the frames and thwarts were lashed to the lugs on the strakes. The similarity of the stem and stern structure of the Punjulharjo ship to other ship remains from Southeast Asia, including the Butuan boats, has been pointed out (Clark et al. 1993; Manguin 2009). Radiocarbon dating of the Punjulharjo ship timbers dates the construction period to AD 660–780 (Priyanto 2011).

Three shipwrecks found in the Java Sea—known as the Intan, Cirebon, and Karawang shipwrecks—are said to date back to around the tenth century, and they can probably be assigned to the category of seagoing ships constructed by techniques endemic to Southeast Asia. The hull remains of the Intan shipwreck were poorly preserved, but salvage reports and studies on cargo are available (Mathers and Flecker 1997; Macleod and Flecker 2001; Flecker 2002; Twitchett and Stargardt 2002; Guy 2004). Few archaeological reports have been produced on the Cirebon and Karawang shipwreck sites, but these shipwrecks appear to have been well preserved; planks having lugs were revealed, and at least some frames remained when the ships were discovered. Another example of a hull structure related to Southeast Asian techniques is a shipwreck discovered in 2010 in the waters off the coast of Tanjung Simpang Mengayau, in the state of Sabah in Malaysia. The remaining part of the hull shows lashed-lug construction (Flecker 2012). Ligaya Lacsina (2011) recently reported the finding of plank remains in the waters of the Philippines that were originally assembled using techniques similar to those visible in the Butuan boats.

The waters off the coast of Borneo were a corridor for Southeast Asian seagoing ships that sailed between the archipelago and the mainland of Southeast Asia. In Brunei, the Brunei Museum Department conducted research at the Sungai Limau Manis site, twenty-two kilometers from Bandar Seri Begawan, and identified numerous ship timbers, including bow parts, hull planks, frames, and paddles (Yahya 2005). While the absolute date of the ship remains has not been ascertained, Chinese ceramics excavated at this trading site, including Yue wares, are dated to the Song period. The Kota Batu site is another important site in the country that involved trade. Its active period has been assigned to the Ming dynasty (Harrison 1976; Omar 1983). Excavations carried out at the site during 1952–53 revealed a few ship construction tools and materials, such as resin, as well as ship timbers (Yahya 2005). The hull remains, however, were not recorded in detail.

Major discoveries of ship remains reflecting the lashed-lug technique are not restricted to the Southeast Asian archipelagos. Pierre-Yves Manguin (1993) reported the discovery of ship remains at the site of the Wat Khlom Tom temple in Krabi Province in southern Thailand. The site is known as a bead production site (Khuan Luk Pat or Kuan Look Pad) active from the fourth to the ninth century, when the area thrived because of maritime trading. The ship remains were found in 1966 in the Tum River vicinity near the temple site, and some timbers were recovered later, in

2012. Some planks have rectangular lugs, and there is a broken piece of timber that could have been used as part of the bow or stern (figure 8.2). The timbers are stored in the Wat Khlom Tom Museum and are not well preserved, but it is still obvious that some timbers are quite distinctive from each other in shape and probably came from different ships. The degradation of the timbers, however, prevents identification of the original configuration of the lugs.

The richness of the lashed-lug construction tradition along the Thai coasts is also evident in ship remains found in the Tran Province, which adjoins the southern part of Krabi. Ship remains were found in 1989 during the dredging of an artificial pool behind the Wat (Kok Yang) temple in Kan Tang district. Among Thai scholars there seems to be an insistence that the ship originated from the Indian Ocean, considering the distinctive features of the ship timbers and the location where the remains were discovered, which used to be part of the Tran River connecting to the Andaman Sea. What is visible in photographs, however, which show the lugs and stitched planks, indicates that the original ship might have been built in the Southeast Asian shipbuilding tradition. Unfortunately, the ship timbers have been destroyed. The lashed-lug technique and stitched planks have also been identified in ship timbers found on the eastern side of Thailand. A few ship remains were discovered during the development of the Nong Ree canal in Koh Kwang village, Muang district, in Chanthaburi Province; they were recorded by the Thai Underwater Archaeology Division, under the Department of Fine Arts.

Widespread use of the lashed-lug technique along the coasts of the islands and mainland of Southeast Asia implies standardization of the technique across the region. Southeast Asian indigenous craft found in the surge of archaeological discoveries are undated, but they illustrate a sphere of technology that covers almost the entire South China Sea. The most recent finding to be noted is on shore alongside Chau Tan village in Binh Son district, Quang Ngai Province in Vietnam. A shipwreck with a large number of Chinese ceramics including Changsha ware, Yue celadon, Ding- and Xing-type white porcelain, and Three Color Ware has been discovered. Some bowls and jars have been described as bearing Chinese, Arabic, or Indic inscriptions (figure 8.2). Yet as is true of many of the above-mentioned cases, the ship remains have not been properly recorded.

Figure 8.2. Ship timbers stored at the Wat Khlom Tom Museum, Krabi Province, Thailand (*top*), and ship timbers found in the waters around Cu Lao Cham in central Vietnam (*bottom*). (Top photo by author; bottom photo by Do Troulg Giang)

Hybridization Theory in the South China Sea

> As noted before, this South China Sea tradition is not a technical family that can be defined in clear-cut terms; it is not a homogenous group, since this tradition clearly bridges national borders and is identifiable in both China-built and Southeast Asia–built vessels. (Manguin 2010: 345)

The South China Sea shipbuilding tradition is characterized by structural uniqueness evident in numerous excavated ships in the region: hybrid features reflecting technologies from both East Asia (southern China) and Southeast Asia (Manguin 1984, 1993, 2010). Innovation in the hull structure and construction methods is represented by the shift of the transverse structure from frames to bulkheads, by the disappearance of the lashed-lug technique, and by the retention of wooden fastenings for the main planking. These hybridization processes are considered examples of an integration into Southeast Asian shipbuilding of technologies attributed to East China Sea shipbuilding.

The rise of East China Sea shipbuilding during the Song dynasty followed the maturing of the shipping business and trade in northern and central China (Shiba 1968). One can argue that sinicization of shipbuilding in the southernmost areas of the Chinese coasts in the early Song dynasty could have lagged behind, as indicated in the historical text *Huilin Yinyi*, quoted earlier. Nevertheless, expansion of maritime networks involving cities in Fujian Province in southern China and connecting to Southeast Asian ports definitely occurred throughout the Song dynasty period (Clark 1991; Ptak 1998; Schottenhammer 1999; Heng 2009). The linkage between the East China Sea and South China Sea was enhanced by the engagement of Chinese traders in long-distance maritime trading.

Shipping between the two seas continuously increased as the Mongol Empire gradually strengthened its control of territories in China. The hegemony of the Yuan, descended from the Mongol Empire, expanded after garnering maritime power by defeating the Southern Song in 1279. The Yuan rulers actively used ship transportation for military campaigns in various places in East and Southeast Asia (Rossabi 1988; Delgado 2009; Kimura et al. 2014). Sending out the fleets affected the power relations of the polities engaged in maritime activities and changed the sailing routes (Ptak 1998). Decline began for Srivijaya, which had dominated intra- and transregional maritime networks in Southeast Asia from the seventh through the fourteenth century, and the subsequent rise of the Majapahit (AD 1293–1527) is a representative case of the impact of armed intervention by the Yuan dynasty (Miksic 1983; Yamamoto 1983). In the end, however, Yuan aggression proved unsuccessful. Beyond the naval campaigns, we should note that Yuan maritime policies had the capacity to promote more seaborne trade than did the policies of previous dynasties, as is also evident in the development of extensive Sino-Muslim networks during that time (Chaffee 2008; Yasuhiro 2008).

Thus, the diffusion of the Chinese shipbuilding tradition was part of the maritime expansion through the Song and Yuan periods. Direct voyages of ships built in China to trade in the South China Sea probably occurred in the twelfth century, and the arrival of Chinese traders in Southeast Asian ports did not cease in the following periods. Strengthening of the relationship between the Yuan dynasty and its tributary states in Southeast Asia probably enhanced the volume of traffic on the sea. All these factors directly bear on the matter of shipyards in the South China Sea regions rapidly adapting technologies from the East China Sea. They started to construct hybridized ships that had never been built in the regions before.

Advent of the South China Sea Shipbuilding Tradition

The timing of the dawn of hybridization in shipbuilding technologies in the South China Sea is controversial, but the increased number of archaeological sites associated with ships that originated from East or Southeast Asia provides some clues for addressing this question. An inventory of ship remains and wreck sites in the waters of the Philippines, Indonesia, and Thailand has been developed (appendix 3). As a first step, the list of ship remains can be used to consider the earliest evidence of seafaring of Chinese ships in South China Sea waters. In the above-mentioned countries, only a few sites have been identified that possibly date to the Song dynasty period of the twelfth or thirteenth century. In the Philippines, some ceramic remains and stone anchor stocks found at reef sites—Bracker Reef, Investigator Shoal, and the Bolinao II site on Silaqui Island—appear to relate to wrecking events that date back to that period (Clark et al. 1989; Dizon 1992). The stone anchor stocks from sites in the Philippines appear to have originated from Chinese ships, being similar to the ones

Figure 8.3. Stone anchor stock from a twelfth- to thirteenth-century wreck site at Investigator Shoal, stored in the National Museum of the Philippines. (Photo by author)

discovered in Quanzhou in China and along the Japanese coast (figure 8.3). While the dates for these sites are based on ceramic assemblages that assigned them to the eleventh and twelfth centuries, further detailed study of the stone anchor stocks may complement perspectives on direct voyages of East China Sea traders into Southeast Asia.

Stone anchor stocks identical to the ones found in the Philippines were discovered at the Jepara wreck site in Indonesia. The wreck site had been heavily pillaged, but the assemblage of ceramics recovered and reported consists of Chinese ceramics from kilns in Fujian, such as Tongan and Dehua (Atmajuana and McKinnon 2005). The recovered ceramics and copper coins suggest that the shipwreck can be dated to the twelfth century. One of the two stone anchor stocks is approximately two and a half meters long, and the middle part of the stock is grooved. Indonesia has a wreck site, known as the Pulau Buaya wreck site, dated to the Song dynasty (Ridho and McKinnon 1998; McKinnon 2001). Cargo from the shipwreck consists of diverse trading commodities: more than 32,000 pieces of ceramics, ingots, copper coins, and glasswares, along with what are probably personal items, including wooden dice. The archaeological report presents intensive information about the cargo, but the condition of the hull remains is not clearly known.

So far, clues to determine the advent of East China Sea traders in the South China Sea before the fourteenth century are extremely limited. While a few wreck sites dating to the twelfth and thirteenth centuries have been reported, almost no information about the hull condition of these wrecked ships is available. Known wreck sites of seagoing traders from fourteenth-century China are more limited in Southeast Asia than are remains from the previous periods. None of the shipwrecks from the Philippines or Indonesia, for example, has been clearly assigned to this period. In Vietnam, local authorities have announced the discovery of the remnants of a fourteenth-century ship lying off the coast of Binh Son district in Quang Ngai Province. The discovery of many ceramics associated with the shipwreck was emphasized in an announcement during salvage operations by a local company using a dry-dock method. Despite the fact that a large portion of the hull was well preserved when discovered, the salvage operation appears to have focused primarily on the recovery of ceramics.

Wreck sites in Thailand have been reasonably fully reported, providing some reliable data about hull structure and construction techniques. Several shipwreck sites in the Gulf of Thailand have been archaeologically surveyed and excavated. Underwater archaeology in Thailand started in the late 1970s, and an initial underwater survey on the east side of the Gulf of Thailand identified twelve wreck sites (Intakosai 1984; Prishanchit 1992; Vatcharangkul 1992). Of these, Ko Rang Kwien is the earliest ship, dating to the late fourteenth or early fifteenth century. The hull was substantially damaged, and only part of the ship survives, including some keel timbers. The keel scarf has holes for the placing of copper coins or a mirror, as was the custom of Chinese shipwrights (as seen in the Song and Yuan dynasty ships described in chapters 4 and 5). Bulkhead planks have not been identified, but the ribs (frames) are fastened into the hull planks by wooden pegs. Whether the hull originally had a bulkhead structure has not been confirmed. Thai researchers have interpreted the Ko Rang Kwien shipwreck as a locally built ship; yet it could have been constructed partly based on the Chinese shipbuilding tradition. The coins and mirror in the keel joint are indicative of the transfer not only of construction skills but also of ritual practices attributed to Chinese shipbuilding. This raises questions about how shipwrights' beliefs and behavior were transferred from one country to another. Is this suggestive of a migration of shipwrights who established their industry on the coasts of Thailand?

Ko Si Chang shipwrecks nos. 1–3 and the Pattaya shipwreck are other cases important for insight into the types of ships in the Gulf of Thailand. These are a group of shipwrecks investigated through a collaborative project by the Thai Underwater Archaeology Division and researchers from the Department of Maritime Archaeology of the Western Australian Museum (Green et al. 1986; Atkinson et al. 1989). Of these shipwrecks, Ko Si Chang no. 2 is the oldest ship, according to radiocarbon dating of a timber, which yielded a calibrated date of AD 1290±60 (Atkinson et al. 1989); a study of recovered ceramics dated the ship to the beginning of the fifteenth century (Mukai 2000). The fact that the absolute dating of construction material shows an earlier date than the relative date of the lading item could be valid, although there is a gap between the two dates. The dates nevertheless classify Ko Si Chang no. 2 to a relatively early period.

Ko Si Chang shipwreck no. 1 has been dated to a later period, according to radiocarbon dating of its dyewood cargo to the late sixteenth century, which corresponds with the date of its cargo of Thai ceramics (Atkinson et al. 1989; Mukai 2000). The hull structure of the two shipwrecks is distinctive. The Chinese shipbuilding tradition appears to have been more influential in the hull structure and construction methods of the older Ko Si Chang shipwreck no. 2. Iron fastenings were reported as having been used for the innermost planks, but these reflect a construction method different from those typically observed on the planking of East Asian ships (Atkinson et al. 1989). Ko Si Chang no. 1 clearly represents the South China Sea shipbuilding tradition, as indicated by its reported planking (Green et al. 1986).

The cargo assemblage for Ko Si Chang shipwrecks nos. 1 and 2 indicates that they were international traders. In contrast, Ko Si Chang no. 3, dated to the mid-sixteenth century, is considered to have been a coastal trader (Green et al. 1987b). The Pattaya shipwreck is likewise considered to have been engaged in coastal trading (Green and Harper 1983; Green and Intakosai 1983); it has been tentatively dated to the mid-sixteenth to mid-seventeenth century. The hull structure of these local traders indicates that shipbuilding industries in the middle and eastern coasts of the Gulf of Thailand had fully adopted the East China Sea shipbuilding tradition early, likely during the fourteenth century.

Gulf Traders

According to Suebsang Promboon (1984), after the Ming dynasty came to power in 1368, Thailand delegated seventy-eight trade missions to China for tribute from 1371 to 1511. The Ming court sent seventeen tributary missions to research Thailand during the same period. In view of the fact that the Ming court restricted private maritime trading, it is interesting to consider whether merchants and traders in Southeast Asia attempted to raise their prominence in the South China Sea maritime trade. A substantial increase in the exportation of Thai ceramics to all regions of Southeast Asia has been reported for the mid-fifteenth century (Mukai 2000). Possible growth of shipbuilding industries along the coasts of the Gulf of Thailand around the same time should be considered. Traders from the Gulf of Thailand could sail widely in the South China Sea and even to the East China Sea. This represents the heyday of gulf traders.

The increase in hybrid types of shipwrecks dating to the fifteenth century onward is notable when appraising shipwrecks found in waters off Thailand. A shipwreck found in Koh Kong Province in Cambodia, for example, has the potential for further comparative study with gulf traders found in Thailand. The Cambodian shipwreck was found about twenty kilometers offshore of Sdach in Kirisakor district (Ministry of Culture and Fine Arts 2011; Phann 2011; Beavan et al. 2012). Recovered ceramics include Thai wares from the kilns of Sukhothai and the river Noi. The ceramic assemblage and radiocarbon dating indicate that the ship dates to the late fifteenth to early sixteenth century (Beavan et al. 2012). The overall assemblage of its cargo—not only ceramics but also other commodities, such as several lead ingots—appears to be very similar to that of Ko Si Chang no. 1. A full underwater excavation at the site in the future may deliver more details on the hull structure.

Vietnamese studies have recently reoriented the evaluation of the kingdom of Dai Viet in a new direction in terms of its role in maritime trading in the South China Sea, suggesting that it represents the establishment of a trading center in northern Vietnam (Momoki 1998; Li 2006). The historical significance of international ports has been pointed out in relation to the expansion of Vietnamese ceramic exports in the fifteenth century, as well as a shipwreck discovered off Hoi An on the south-central coast of Vietnam (Abe 2010). The assemblage of recovered Vietnamese and

Chinese ceramics and radiocarbon dating of a ship timber date the Hoi An shipwreck to the late fifteenth century, and its hull planks appear to be composed of woods native to Thailand (Kikuchi 2010). Many ships built along the Thai coast presumably sailed into the South China Sea off the Vietnamese coast, but whether the Hoi An shipwreck originated from the Gulf of Thailand is still open to question. Shipbuilders in Hoi An could have developed the capacity to supply hybrid ships in later periods. Hoi An, on the south-central coast of Vietnam, grew as an international trading port under the reign of the Nguyen lords (AD 1558–1777), who placed a foreign settlement in Hoi An. The Nguyen lords also promoted the industrialization of shipbuilding through various preferential incentives and established government shipyards with hierarchical employment of skilled shipwrights (Nishimura and Ueda 2012). These policies facilitated the passage of Dutch, Chinese, and Japanese merchants into Hoi An during the seventeenth century. Hoi An was one of the destinations for the red (or vermilion) seal vessels (*shuinsen*) of Japanese merchants, which undertook long-distance voyages between the East China Sea and South China Sea; some of the shuinsen display hybrid features (see chapter 1).

The Philippines, which lie on the margin of the South China Sea, have been a stopping point because of their proximity to East Asia. The Santa Cruz shipwreck discovered in the Philippines serves as an example of a fifteenth-century trader found away from the Southeast Asian mainland. Data on its cargo and hull structure have been reported (Orillaneda 2008). The site was discovered offshore from the municipality of Santa Cruz in Zambales Province. The Santa Cruz shipwreck is a late fifteenth- to early sixteenth-century merchant ship, according to Bobby C. Orillaneda's study on the assemblage of ceramics loaded as cargo. Its hull measures approximately twenty-five meters long and six meters wide. The remaining parts include a keel, sixteen bulkheads, and half frames. Eleven or twelve strakes appear to remain. The hull strakes are edge-joined, showing three-layered planks. The planking has wooden dowels fastening the innermost planks and iron nails fastening the inner to the outer planking. The hull has bulkheads as its main transverse structure. The main structure of the hull uses Chinese construction methods, but planking using wooden dowels for the inner planks is a technique endemic to Southeast Asia. The results of wood species identifications on the timbers suggest that the Santa Cruz ship was locally built (Orillaneda 2008).

Southeast Asian Nautical Archaeology in the Asian Century

Despite the increase in the discovery of shipwrecks over in the past few decades, Southeast Asian maritime archaeology provides limited information for the study of hull structure and construction methods. This is partly because of a lack of resources for local researchers to conduct detailed investigations of hull remains. An exception is the Thai Underwater Archaeology Division, which has been actively carrying out underwater archaeological expeditions after ceasing collaborative projects with overseas experts (Green and Harper 1982). Collaboration led to local capacity building for extensive surveys and excavations guided by local strategy. One of the remarkable outcomes of this endeavor has been the excavation of the Bangkachai II shipwreck. An internal site report in the Thai language is the only data currently available.

The Bangkachai II shipwreck was discovered in 1992 in eight meters of water in Bangkachai Bay, approximately two kilometers from the coast of the Bangkachai community in Chantaburi Province. The hull bottom remained in good condition when discovered. The remaining part measures approximately twenty-four meters in length and six meters in width. The hull bottom cross section shows a rounded shape. The remaining bottom structure includes a keel and thick garboards that are joined together with wooden fastenings. Hog pieces reinforce the joint of the keel members. At one of the joints between the keel members, evidence of *bao shou kong* has been identified. Longitudinal timbers run from the bow to the stern, penetrating the bulkheads. Nineteen bulkheads remain. The lower and upper bulkhead planks differ in thickness. Most of these planks are joined by wooden fastenings, but in some parts skew-driven iron nails are used in both the forward and aft faces of the bulkheads. The bulkheads have triangular limber holes. A bamboo stick was found next to a limber hole; it had probably been used to clear the way for bilgewater, not to seal the limber holes (Worrawit Hassapak, pers. comm., 2009).

Twelve strakes remain on the port side, and nine strakes remain on the starboard side. The hull planking has sheathing planks forming the two outer layers. The width of the planks measures 200–250 millimeters, and the thickness varies from 50 to 80 millimeters for the inner planks and from 40 to 60 millimeters for the outer planks. The strakes are edge-joined, and the planking uses both wooden and iron fastenings. Wooden dowels (treenails) seem to have been identified for the butt joint of the

Figure 8.4. Bangkachai II shipwreck, scale model. (Courtesy of Thai Underwater Archaeology Division; photo by author)

strakes. Two mast steps were found; the forward mast step measures 2.98 meters athwartships, and the mainmast step measures 3.18 meters athwartships. Two recesses are cut to hold the heels of the mast cheeks (figure 8.4).

Three large wooden anchors were recovered during underwater operations. The main hull components were made of tropical woods, identified as *Cotylelobium lanceolatum* Craib and *Tectona grandis* L.f. (commonly known as teak). The wood species indicate that the ship was probably built in Southeast Asia. Radiocarbon dating yielded a year of 360±70 BP, and the approximate date of the wrecking is estimated at around the 1610s, given the results of analysis and cross-dating based on Chinese ceramics dated to the Ming dynasty's Wanli reign (AD 1573–1619) and Chinese inscriptions on a wooden box storing small lead ingots.

The Bangkachai II shipwreck is regarded as a merchant ship, as illustrated by the assemblage of trading materials. The main cargo was probably a large amount of sappan wood for dye (approximately 20,000 pieces) and Burmese ironwood (*Xylia xylocarpa*). Other plant products, such as *Terminalia* nuts, betel nuts, and peppers, were also present. Many copper ingots (totaling ten tons) of different shapes were found. The quantity of ceramics among the artifacts is low, indicating that they were not major trading materials or perhaps were personal belongings. Some jars contain tamarinds, peppers, and animal and fish bones. Some of the stoneware and earthenware is from Thailand's Noi River, in Sangbui Province, and some Sangkalok wares, such as green glazed ceramics, are from the Sukhothai period. The ceramics discovered include overseas products from Vietnam, though notably the relative dating on the Vietnamese ceramics shows them as from the early sixteenth century, approximately one hundred years older than the dates given for the ship.

Part of the hull and its outfitting—such as a mast step, anchors, and cannons—have been recovered. Conservation work will be needed on these recovered components. The hull is preserved in situ. I am not aware of the monitoring and management strategy for the hull, but it constitutes a significant case for developing appropriate in situ preservation methods in the region. In many cases in Southeast Asian waters, including the South China Sea, past operations for commercial salvage at wreck sites have focused on artifact recovery, especially of ceramics, and have given recording of the shipwreck hull low priority.

Although some notable recording projects have been conducted, the original data are dominated mostly by salvage companies and have not necessarily been shared with local people or the countries responsible for custody of sites to make data available for further study. At the end of a salvage operation, the site has frequently been left without management or protection policies; rescue of artifacts has been expedient in past salvage operations, rather than a vision of protecting the hull as an archaeological site. Knowledge of wreck sites in the South China Sea is still scattered and lacks broad perspectives, despite the fact that in many cases the wreck sites have a transnational historical background, not only in the assemblage of seaborne commodities but also in hull structure and construction methods. As demonstrated in this chapter, there is a need for a regional perspective in the study of shipwrecks in the South China Sea.

9

Conclusion

My broad aim in this book is to clarify the historical development of shipbuilding technologies in East Asia over the centuries by assessing them at a regional level and from the viewpoint of maritime archaeology. As previously discussed, the analysis is based on an approach centered on the concept of hybridization underlying innovations in shipbuilding, as well as hypothetically considering that evidence for technological hybridization is ubiquitous and can be found much more widely on excavated ships than was previously thought. By taking this approach, I intend to reorient in a new direction the existing linear theory of evolution for the interpretation of shipbuilding and naval history of one country. The results of intensive and extensive study of multiple ships excavated in East and Southeast Asia verify the conceptual frameworks. Detailed explanations of structure and construction methods observed in the ship remains, especially with a focus on representative hull components (such as bulkheads and shipbuilding materials) allow for a comparative assessment based primarily on archaeological data but also multidisciplinary in nature, drawing data from ethnohistorical, textual, and experimental sources as well.

Summary of East Asian Shipbuilding Technologies

East Asian shipbuilding technologies were adopted variously to form regional traditions with distinctive characteristics. Among these were dominant and lasting technologies that are evident along the coasts of the Yellow Sea and the rise of the East China Sea shipbuilding tradition, dissemination of which influenced the formation of the South China Sea shipbuilding tradition.

Yellow Sea Shipbuilding Tradition

Technologies are evidenced in three different types of ships: (1) excavated Tang dynasty riverine ships dating from before the tenth century that have a flat bottom and use bulkheads and iron fastenings; (2) excavated Goryeo dynasty coastal traders of the eleventh to fourteenth centuries off the western coast of the Korean Peninsula that have a hull floor with robust chine timbers, use beams for transverse components, and have wooden fastenings; and (3) excavated flat-bottomed keelless ships of the Ming dynasty that were constructed using shipbuilding technologies of the Korean Peninsula but were hybridized with Chinese shipbuilding technologies, as illustrated by the use of bulkheads and iron fastenings. Shipbuilding in the Korean Peninsula had developed endemic technologies at least as early as the tenth century, and these technologies were influential outside the Yellow Sea, such as in Japan.

At the same time, another major type of watercraft sailed between the southern regions of Japan and northern coasts of China. These ships were probably similar to Chinese coastal ships used from Hangzhou Bay northward, areas that formed a sphere of construction for flat-bottomed ships. The water environment as well as the development of riverine infrastructure in these areas facilitated the advent and lasting use of such ships.

East China Sea Shipbuilding Tradition

Details of early periods of shipbuilding in this tradition are still disputable; however, the middle coast of China is considered to be the northern limit of the range of the Austronesian shipbuilding tradition, and various groups of seafarers reached this coast before the tenth century. Endogenous growth and exogenous factors combined to promote further shipbuilding development in this area. Technologies in later periods are evidenced in two types of ships. The first set consists of excavated seagoing ships of the thirteenth and fourteenth centuries that have a deep deadrise (a V-shaped bottom in cross section) formed by a keel and garboards. The transverse structure consists principally of bulkheads and is further secured by half frames and brackets. The hull is constructed of layered planking and uses various types of iron fastenings. The second set comprises excavated ships dating from the Song to the Ming dynasty with a rounded bottom (a U-shaped bottom in cross section) and a keel. The hull is likely to have been constructed by single-layered planking (baulks) and has bulkheads.

The technologies of the first set (keel, garboards, bulkheads, half frames, brackets, layered planking, and iron fastenings) show a strong connection with the southern part of the Chinese coast around the Taiwan Strait and southward. In contrast, the focal area of the second set of technologies can be identified to the north, around Ningbo, but this technological suite is regarded to have become dominant in the broad area of the Chinese coasts than the first one in the later periods. Moreover, shipwrights using this technology of the East China Sea shipbuilding tradition could have had a mutual interaction with those of the Yellow Sea shipbuilding tradition, as indicated by identical ships found throughout the two regions.

South China Sea Shipbuilding Tradition

The South China Sea region, which includes the southern coast of China, was dominated by Southeast Asian indigenous shipbuilding traditions. Wooden fastenings (treenails, or dowels) and lashed or stitched planks represent the typical planking technology used in ships in the region. Technological advancement occurred early in the first millennium AD with the advent of the lashed-lug technique, which brought the installation of frames and thwarts into the hull, thereby promoting the construction of larger seaworthy ships. That lashed-lug construction was a prevalent technique is illustrated in how widely such ship remains have been found: both on land and in waters from the Malay and Indochinese peninsulas across insular Southeast Asia.

The prominence of the technique, however, was probably diminished by the more active involvement of ships from the East China Sea around the late thirteenth to the early fourteenth century. Shipbuilders in some zones of the South China Sea, such as the coasts of the Gulf of Thailand, innovated through hybridization by integrating the existing and exogenous technologies. Hybridized features are evidenced in shipwrecks dated to the fifteenth century and later that are found in many waters in Southeast Asia. They have a keel and a round bottom. The transverse components consist of bulkheads and, likely, half frames. The hull planking is in multiple layers, and the main planking (that is, the innermost planking) has wooden fastenings with compensatory uses of iron fastenings.

Toward a New Interpretation of East Asian Shipbuilding Technologies

An increase in the number of archaeological discoveries of coastal traders and seagoing ships in East Asia during the last quarter of the twentieth century has permitted multiple excavated ships identified in China, Korea, and Japan to be analyzed using a comparative perspective at a regional level. Such an approach has not been attempted in prior scholarship, and it shifts the emphasis toward analyzing the dynamics of regional shipbuilding rather than that of a single nation. This macro view should be carried forward into later East Asian shipwreck archaeology, so that comparative research on structural features and construction methods of the excavated hulls can be given in as much detail as in nautical archaeological research on ships found in Europe and the United States.

My research centers on the examination of representative hull components of seagoing ships, such as the bulkhead, a common structural feature found in trading ships that were engaged in maritime activities all across East Asia. The bulkhead is defined as an indispensable hull component for long-distance voyages and provides a schematic concept for East Asian seagoing ships. In this sense, this transverse component can be used as a parameter for evaluating the formation of regional shipbuilding traditions.

The development of regional shipbuilding technologies in East Asian shipbuilding history was dependent on technological innovations, particularly those that involved the hybridization of ship construction methods. In particular, the integration of specific hull components was directly related to the formation of regional shipbuilding traditions. While previous studies presented technological hybridization as localized, my comparative research presumes that the identification of hybridized features can be more readily observed through a wider approach to the formulation of regional shipbuilding development in both East Asia and Southeast Asia.

Further, this research shows the value of multidisciplinary study in East Asian maritime archaeology by incorporating both archaeological data and various other kinds of data, including historical accounts and iconographic and ethnological resources. The multidisciplinary review in chapter 3 integrates into a single account the recent work on these resources by regional and extraregional scholars. The review reveals the availability of secondary data from historical and boat ethnological studies on East Asian maritime trading, seafaring, and traditional boatbuilding. With regard to archaeological data, chapter 3 also clarifies achievements in underwater archaeological survey and excavation in China, Korea, and Japan. Archaeological reports from these operations are used as primary data in this research.

Chronology of Principles in East Asian Shipbuilding Traditions

Representative hull structures and construction methods of excavated ships in East Asia (in particular, those related to the three shipbuilding traditions discussed throughout this volume) are specific to each body of water: the Yellow Sea (see chapters 4 and 5), the East China Sea (see chapters 6 and 7), and the South China Sea (see chapter 8). Archaeological evidence from the detailed study of hulls and ship construction materials yields insights into the emergence of representative technologies, which can thereby be incorporated into the dynamics of East Asian shipbuilding.

Early Features of Shipbuilding in the Yellow Sea

The development of the flat-bottomed ship, a shipbuilding tradition endemic to the Korean Peninsula, is one of the ancestral threads of East Asian shipbuilding. It had a notable impact on the prominence of coastal traders during the Goryeo dynasty (AD 918–1392) and may have influenced early Japanese shipbuilding technologies. Its possible linkage with Japanese shipbuilding is posited in this volume for the first time. While ship remains confirming the hypothesis have not yet been found in Japan, there is nevertheless room to reconsider the existing linear theory that has been advocated for the evolution of traditional Japanese ships. A fourteenth-century historical account defines traditional Japanese ships as flat-bottomed ships. Furthermore, some representative features of the major type of coastal traders used from the seventeenth century onward, recognized as a flat-bottomed ship, are akin to the structure of early coastal traders used in Korea.

Nine hulls dating to the Goryeo dynasty have been excavated on the western side of South Korea. They were constructed using the same shipbuilding technology, an endogenous form presumably developed before the Goryeo dynasty, although all the excavated ships are dated to the early eleventh century onward. Representative features are baulks as longitudinal structures in a flat-bottomed hull, athwart beams as transverse structures, and wooden fastenings. These Goryeo coastal traders were designed to sail the muddy and shallow waters of the Yellow Sea. They illustrate the dominance of seaborne activities by merchants from the Korean Peninsula in the early East Asian waters, as mentioned in historical records.

The long-term dominance of flat-bottomed ships traditionally used in northern waters of the Chinese coast is evident in various resources (see chapter 2). Evidence from boat remains in China verifies the construction of planked boats with a flat bottom before the tenth century, around the Tang dynasty period (AD 618–907). Many boats have been found in or near ancient waterways or canals in northern China. Two Tang dynasty boats were discovered inland in Jiangsu Province. All the boats that have been dated to this period are considered to have been for riverine use. Bulkheads, which are regarded as a representative characteristic of historical Chinese ships, initially appeared in the structure of the riverine ships prior to their adoption in seagoing ships. They could have developed initially for transverse strengthening of the hull of riverine boats rather than as watertight compartments for seagoing ships.

Ships similar to the riverine boats—with a flat-bottomed hull and bulkheads—may have been used in early China for sailing out to sea. People around the coast of the Yellow Sea probably pioneered the use of flat-bottomed ships at sea. Jiangsu Province and Zhejiang Province to the south of it (see figure 1.1) are described as having had mother ports for ships conducting distant voyages since before the tenth century. These were focal ports for many envoys and trading ships involved in early maritime interactions between the Chinese continent and Japan during the Tang dynasty. While we can only deduce the possible structure of those ships crossing the Yellow Sea and the East China Sea, there likely were not substantial differences in the hulls of early Chinese ships for riverine use and for seafaring. Even in the nineteenth century, the *sha-chuan* (sand ships), represented by a Hangzhou Bay trader, were still perceived as ships used for both riverine and seagoing purposes in the Yellow Sea and the northern East China Sea (see chapter 3). Limited evidence allows us to infer that during the early stages of shipbuilding along the northern coasts of China, the structure and construction methods of seagoing ships in this region were fundamentally identical to those of the excavated riverine ships.

Thirteenth- and Fourteenth-Century Development in the East China Sea

The second sphere of shipbuilding traditions is evident in ships that originated from the East China Sea. We still lack direct evidence for the formation of the area's shipbuilding traditions and for the timing and locations involved. The central and southern Chinese coasts should be regarded a focal region where the transition from riverine to seagoing vessels presumably occurred during the Song dynasty (AD 960–1279). The embryonic East China Sea shipbuilding tradition probably lay in the coastal areas southward around Hangzhou Bay, where shipbuilding industries made a swift switch toward supplying seagoing ships.

Two excavated ships are analyzed in detail to examine the East China Sea shipbuilding tradition: the Quanzhou ship, dating to the second half of the thirteenth century (see chapter 4), and the Shinan shipwreck, from the first quarter of the fourteenth century (see chapter 5). They evidence standardization of the structure of seagoing ships in these periods. Structural features and construction methods identified in the ships reveal one stage of the process of technological integration. The use of a keel with bulkheads occurs in these two significant discoveries of Chinese traders (the Quanzhou ship is the first seagoing ship discovered, and the Shinan ship is the first Chinese trader discovered in water). Whereas how the keel structure began to be adapted in East China Sea shipbuilding has yet to be clarified, the stage of dissemination of this structure can be considered as a period of innovation. One may presume that the innovation facilitated the construction of seaworthy Chinese traders, which led to the prominence of merchants voyaging from the mainland of China in the waters of East Asia as well as Southeast Asia.

The Quanzhou and Shinan ships are East China Sea traders for a long-distance voyage. Their hulls are characterized by a V-shaped bottom, a shallow draft, and wide beams. The standardized structural features include the keel and hull planking, which contribute synergistically to longitudinal strength. The keel is composed of three members, and thick garboards are fixed to it. The hull planking that extends from the garboards forms a steep deadrise angle. Hull planking is edge-fastened by skew nailing with rabbeted seams. The system of multiple-layered planking has also been standardized for protection of the innermost planks and functions as a plank shell. The bulkheads are an indispensable hull component, providing transverse strength, and they are typically used with large half frames and brackets.

Detailed analysis of the hulls of the two ships identifies some differences. In general, the structure of the Shinan ship appears less sturdy than that of the Quanzhou ship, as indicated in the methods of layering the planking. The layered planking of the Quanzhou ship contributes both to protection of the inner planks and to an increase in the longitudinal strength of the hull. In the Shinan ship, the original intention for the layered planking was for it to function as sheathing that protected the inner planks against marine borers, not to increase the longitudinal strength of the hull. The possible structural weakness arising from this may have caused the distortion of the Shinan shipwreck hull that is evident in the now-hogged keel; reconstruction of the ship lines indicates that the hogged keel was not an original feature. With regard to transverse components, bulkheads in the Shinan ship are fewer in number than in the Quanzhou ship, an arrangement that could have allowed the hull to stow cargo of relatively large size. As noted, however, it may have produced frailty in terms of transverse strength. In previous studies, these ships were individually assessed. The intensive joint analysis of the two ships in the present volume argues for the significance of comparative assessment of structural details and construction methods.

Analysis of Shipbuilding Materials

The comparative approach, when extended to the archaeological examination of ship construction materials such as fastenings and timbers (see chapter 6) contributes to a better understanding of the East China Sea shipbuilding tradition. The analysis of specimens obtained from the hull remains of the Quanzhou ship and the Shinan shipwreck is complemented by data on ship remains from the Takashima underwater site, a unique data set considering the origin of the vessels: the remains of ships the Mongol Empire requisitioned from different regions to compose a fleet. Ship construction materials also yield insight into the industries that provided them.

The remains of nails from the Shinan shipwreck and the Takashima underwater site, examined comparatively and assessed using a metallurgical approach, provided results relevant to a regional perspective. The CT scan and X-ray images of the remains of iron nails from the Shinan shipwreck have revealed that their configuration is skewed. The microscopic cross-section image indicates that manufacturing processes for the nails used in the Shinan ship were similar to those of the iron nails from the Takashima underwater site. The first step was the casting of thin bar iron ingots with a square cross section, and then the iron in the ingots was improved—made softer and more pliable—through a process of decarbonization that yielded iron of a quality suitable for making ship nails. Annealing work for the decarbonization was probably conducted in iron workshops built in shipyards. There is a need for future study of iron remains from archaeological sites of shipyards. Such an inclusive approach can help establish a more comprehensive picture of the long-term use of iron fastenings in East Asian shipbuilding traditions.

The assemblage of wood species used in hull components provides another suite of regional insights. Wood species identifications conducted on samples from the Quanzhou ship and compared to existing data on wood species in the Shinan shipwreck and at the Takashima underwater site revealed a pattern of wood usage. Three major species are commonly seen in ships built in the East China Sea shipbuilding tradition: *Pinus massoniana* Lamb., known as Chinese pine; *Cunninghamia lanceolata* (Lamb.) Hook., or Chinese fir; and *Cinnamomum camphora* (L.) J. Presl, the camphor tree.

Cinnamomum camphora and *Cunninghamia lanceolata* make up the majority of the assemblage among the wide range of woods from the Takashima underwater site. *Pinus massoniana* is the dominant species used for the hull of the Shinan shipwreck. The wood assemblage of the Quanzhou ship consists of all three species and indicates that a specific type of wood was used for each hull component. The keel, for instance, consists of *Pinus massoniana* and *Cinnamomum camphora,* which are harder than the *Cunninghamia lanceolata* used for hull planks. *Cunninghamia lanceolata* was used for sheathing planks of the Shinan shipwreck, probably because the wood is softer and lighter and relatively easier to cut and work. It also has reasonable resistance against marine borers.

A comparison of these data with data on wood species identified in other excavated ships in East Asia, including the Goryeo dynasty ships, reveals differences in the types of wood between Chinese shipbuilding traditions and those of the Korean Peninsula. Archaeological study of wood species complements historical texts, which offer limited precision in indicating which woods were used for shipbuilding. Implementation of wood identifications on the remains of ship timbers testifies to the use of particular types of wood corresponding with specific regional shipbuilding traditions. This probably occurred in relation to the growth of networks between timber industries and shipbuilding industries during the Song dynasty. These industries flourished synergistically at some ports along the coast of the East China Sea.

Expansion from the Fourteenth Century Onward

The East China Sea shipbuilding tradition was continuously growing and represents diversity in Chinese shipbuilding traditions. Some ships excavated in the region clearly show differences in structural characteristics from the twelfth- and thirteenth-century ships. An example of such distinctiveness can be seen in some ships excavated along the coast of the East China Sea, which do not have multiple-layered planking and whose hulls have a less steep deadrise in cross section than do the V-bottomed ships. Some ships excavated around Ningbo, including the Ningbo ship and the Xiangshan Ming dynasty shipwreck (chapter 7), typically show such characteristics. During the Ming dynasty (AD 1368–1644), such diversity could have expanded within the region.

Notably, the structure and configuration of the Xiangshan Ming dynasty shipwreck are similar to those of the Penglai ships of the Ming dynasty that were found in Shandong Province, facing the Yellow Sea. Four ships found at the city of Penglai on the coast of the Yellow Sea are classified into two categories. In terms of construction methods, the characteristics of Penglai ships nos. 1 and 2, dated to the late Yuan to early Ming periods, show attributes of the East China Sea shipbuilding tradition. Their planking methods are similar to those of the Xiangshan Ming dynasty shipwreck. All these ships have longitudinally longer hulls in length/beam ratio than do the hulls assigned to previous periods. The features of the other two Penglai ships show a linkage with the Yellow Sea shipbuilding tradition associated with the Korean Peninsula. A hybrid feature is clearly presented in one Penglai ship, no. 3: its transverse structure consists of bulkheads instead of the transverse beams typically present in the Korean ships dated to the Goryeo dynasty. What underlies the hybridization in the Yellow Sea is probably the fact that the East China Sea shipbuilding tradition became influential around the fourteenth century and later.

The South China Sea is considered to be a third sphere of shipbuilding tradition that arose in relationship with the other two regional traditions. The most definitive technological hybridization followed the expansion of the East China Sea shipbuilding tradition into Southeast Asia. Shipwrecks excavated in Southeast Asian waters, dating to the fifteenth century and later, reveal the adoption of technologies originally from China. The South China Sea has been a focal area for technological hybridization throughout its long maritime history. The excavated seagoing ships in the South China Sea are identical to those traders with distinctive hybrid features recorded in Japanese historical accounts and iconographic resources. From these sources, we can presume that there was a certain demand for the hybrid type of ship.

Behind this lay the recognition and adaptation of innovations associated with the hybrid ships—such as enhancement of their performance at sea and their cargo capacity—among some of the people engaged in shipbuilding around the South China Sea. Shipbuilding in Thailand, for example, experienced substantial growth as a supplier of hybrid ships. Shipbuilding industries in the South China Sea met the West in the fifteenth century; for instance, the Dutch East India Company placed an order with shipbuilders in Ayutthaya for "Chinese-type junks" that would be employed sailing to Japan. What are called Chinese-type junks in historical accounts were probably vessels constructed in the South China Sea shipbuilding tradition. In the late sixteenth to early seventeenth century, Japanese merchants made more-direct inroads into maritime trade with Southeast Asia by sailing the hybrid ships.

While the framework of the South China Sea shipbuilding tradition was not made clear in previous studies, the present extensive research makes clear that the development of this shipbuilding tradition is related to the formation and expansion of the East China Sea shipbuilding tradition. In this regard, the detailed pursuit of hull structure and construction methods has substantial significance, linking shipbuilding traditions over these waters and locating them in a long-term chronological framework.

Appendix 1

Contents of Discovered Materials from the Quanzhou Ship

Type of material		Hold number													At keel joints (*bao shou kong*)	Total (approx.)
		1	2	3	4	5	6	7	8	9	10	11	12	13 (aftmost hold)		
Aromatic material/ medicine	Aromatic products: fragrant wood incense (絳真香), sandalwood (檀香), gaharuwood (沈香)	Y	Y	Y	Y	Y	Y	Y	Y	Y	Y	Y	Y			2,350 kg
	Pepper, *Fructus piperis*	Y	Y	Y	Y	Y	Y	Y	Y	Y	Y	Y	Y			7.5 kg
	Betel nut, *Areca catechu*			4			13			2	12	1	9	10		51
	Olibanum, or frankincense		Y	Y			Y	Y		Y	Y			Y		6.3 g
	Ambergris		Y	Y			Y	Y		Y	Y			Y		1.1 g
	Cinnabar			Y	Y											4.6 g
	Mercury			Y	Y											385 g
	Tortoiseshell, *Eretmochelys imbricata*												1			1
Wooden strips	Label/tag	7	3	3		12	43	1		1	12		7	5		94
Coins	Copper coin	10	15	122	14	4	101	128	37	5	5	4	7	37	13	504
	Iron coin														7	7
Ceramics	Pottery	11	4				3	2	1	4		4	1	4		34
	Porcelain	3		2		1	1	1	1		1	10		4		24
Metal objects	Unknown copper			1												1
	Copper spoon			1												1

	Copper wire												2	1		3
	Copper hook													1		1
	Copper mirror										1				2	2
	Copper chain															1
	Hand ax												1			1
	Iron hook	1														1
	Iron punch												1			1
Wooden material	Bamboo ruler												1			1
	Mallet						1									1
	Scraper											1				1
	Container			1												1
	Pan lid						1									1
	Mast step reinforcement		2	2												4
	Wedge				2		2	2								6
	Chopsticks												2			2
	Lid	4	5	6			10		1	39		6	13			82
	Seal												1			1
	Handle											1				1
	Cork											2	2			4
	Hat						1			6						7
Woven material	Bamboo sheet		1	4		1	2	2		2						12
	Latticework sheet			Y							Y		Y			
	Rope	1	12	6			6	1		4	3	1				34
Leather	Unknown product			1												1
Cultural objects	Ivory chess piece			1			1				1	1	4	13		21

(*continued*)

Appendix 1—*Continued*

Type of material		Hold number													At keel joints (*bao shou kong*)	Total (approx.)
		1	2	3	4	5	6	7	8	9	10	11	12	13 (aftmost hold)		
	Piece of paper												1			1
Accessories	Coral bead							1								1
	Crystal bead												1			1
Seeds/nuts	Coconut shell		1	2			7					3		1		14
	Peach seed	1	1	1								1	19	30		53
	Prune seed												1	1		2
	Plum seed												3	2		5
	Coconut seed						1							7		8
	Ginkgo nut												1			1
	Litchi seed			1						4			6			11
Shells	Clam shell	11	17	50	48		12			1000			69	1079		2,286
	Spiral shell			3			2	7		1						13
	Coral															14.9 g
Bones	Wild boar	4		5		3		1	3					3		19
	Sheep								3	5						8
	Dog		2													2
	Rat		23	2			1		2	7						38
	Fish/bird								1	3			3	2		9
Other	Crystal	Y	Y		Y											2.52 g
	Ashed scale?									Y						45 g
	Unknown yellow substance		Y													0.03 g

Appendix 2

Excavated South China Sea Seafarers' Ships with Lashed-Lug Technique

Country	Site	Place of discovery	Date	Stitches/wooden dowels (pegs)	Shape of lug
Malaysia	Pontian boat	Kuala Pontian, Pahang	AD 260–430	Stitched planks and wooden dowels (treenails) for reinforcement of the planking	Round
	Jenderam Hilir	Jenderam Hilir	AD 465–655	Wooden dowels (treenails)	N/A
	Jade Dragon	In the waters off northern Borneo, Sabah	14th century	Wooden dowels	Small, rectangular
Indonesia	Kolam Pinisi	Kolam Pinisi, near the river Musi, Palembang, South Sumatra	AD 434–631	Stitched planks	N/A
	Sambirejo	Sambirejo, near the river Musi, Palembang, South Sumatra	AD 610–775	Wooden dowels (treenails) and stitches for reinforcement of the planking	Rectangular
	Paya Pasir	Medan, North Sumatra	12th–14th centuries	Wooden dowels (treenails)	N/A

(*continued*)

Country	Site	Place of discovery	Date	Stitches/wooden dowels (pegs)	Shape of lug
	Cirebon shipwreck	In the waters off Cirebon Island, West Java	10th century (?)	Wooden dowels (treenails) and stitches for reinforcement of the planking (?)	Rectangular
	Karawang shipwreck	West Java Sea	10th century (?)	Wooden dowels (treenails) and stitches for reinforcement of the planking (?)	Rectangular
	Intan shipwreck	West Java Sea	10th century	Wooden dowels	N/A
Brunei	Sungai Limau Manis	Bandar Seri Begawan	10th–14th centuries (?)	Wooden dowels (treenails)	Rectangular
	Kota Batu	Mouth of Brunei River	14th century	N/A	N/A
Philippines	Butuan boat nos. 1, 2, and 5	Butuan city, Northern Mindanao	AD 320, 1250, and 1215	Wooden dowels (treenails)	Rectangular
	Gujangan wreck	In the waters off Gujangan Island, Sulu Province	15th–16th centuries	Wooden dowels (treenails)	Rectangular (?)
Vietnam	Changsha ceramic wreck site	In the waters off Quang Ngai	9th–10th centuries (?)	N/A	Rectangular
Thailand	Wat Khlom Tom	Tum River near the Khuan Luk Pat site, Krabi	N/A	Stitches	Round
	Wat Khlom Tom	Tum River near the Khuan Luk Pat site, Krabi	N/A	Wooden dowels (treenails) and stitches for reinforcement of the planking	Rectangular
	Wat (Kok Yang) temple	Kok Yang Subdistrict, Kantang, Tran	N/A	N/A	Round
	Chanthaburi	Nong Ree Canal, Silapornville Housing Project, Koh Kwang village, Muang District, Chanthaburi	N/A	Wooden dowels (treenails) and stitches for reinforcement of the planking	Rectangular

Appendix 3

List of Traders Found in the South China Sea Regions

Site	Place of discovery	Date	Origin	Status of hull remains	Preservation status	Year discovered	Salvaged	Surveyed	Excavated	Purpose
Philippines										
Breaker Reef wreck	Breaker Reef, Palawan	N/A	Song dynasty?	No hull remains	N/A	1991	Yes	No	Yes	Trader?
Investigator Shoal wreck	Investigator Shoal, Palawan	N/A	Yuan dynasty	No hull remains	N/A	1990	Yes	No	Yes	Trader?
Lena Shoal wreck	Lena Shoal, Busuanga, Palawan	N/A	Ming dynasty	High	Nontreatment	1997	Yes	No	Yes	Trader
Marinduque wreck (junk)	Gaspar Island, Gasan, Marinduque	N/A	Ming dynasty? or Siam origin?	Low	Nontreatment	1980	Yes	No	Yes	Trader
Pandanan wreck	Pandanan Island, Palawan	15th century	South China Sea/Vietnamese?	No hull remains	N/A	1993	Yes	Yes	Yes	Trader
Puerto Galera wreck	Puerto Galera, Mindoro	N/A	South China Sea?	No hull remains	N/A	1983	Yes	No	Yes	Trader
Santa Cruz wreck	Santa Cruz, Zambales	Late 15th century	South China Sea	High	Nontreatment	2001	Yes	No	Yes	Trader
San Isidro wreck	San Isidro, Zambales	16th century	Southeast Asia?	Low	Nontreatment	1996	Yes	No	Yes	Trader

Thailand										
Ko Rang Kwien Shipwreck	About 800 m north from Rang Kwien Island, west of Bang Sarea Bay	14th–early 15th century	N/A	Low	Nontreatment	1978	N/A	Yes	Yes	Oceangoing trader
Bang Sak shipwreck	1.6 km from Thabtawan Beach, Phang Nga	Radiocarbon date of 760±220 BP	N/A	Low	Nontreatment	2008	N/A	Yes	No	N/A
Ko Si Chang 1	Ko Sichang, Chonburi	Late 16th century	N/A	Low	Nontreatment	1982	N/A	Yes	Yes	Oceangoing trader
Ko Si Chang 2	Ko Sichang, Chonburi	14th–15th century	South China Sea	Low	Nontreatment	1982	N/A	Yes	Yes	Oceangoing trader
Ko Si Chang 3	Ko Sichang, Chonburi	Mid-16th century	Siamese?	Low	Nontreatment	1985	N/A	Yes	Yes	Coastal trader
Ko Khram or Sattahip shipwreck	Near Ko Kram Island, which faces Sattahip Bay, Chonburi	1450–75	N/A	Low	Nontreatment	1974	N/A	Yes	Yes	Oceangoing trader
Klang Aow 2	Middle of Gulf of Thailand, 100 km from Sattahip, Chonburi	16th century	N/A		Nontreatment	2004	N/A	Yes	No	N/A
Hin Bush wreck site	Rayong Province?	16th century	N/A	Low	Nontreatment		N/A	Yes	Yes	Oceangoing Trade
Ko Kra	14 km from Ko Kra, 94 km from Pak Panang	16th century	N/A		Nontreatment	2005	N/A	Yes	No	Oceangoing trader

(*continued*)

Appendix 3—*Continued*

Site	Place of discovery	Date	Origin	Status of hull remains	Preservation status	Year discovered	Salvaged	Surveyed	Excavated	Purpose
Ko Tao	N/A	16th century	N/A	Low	Nontreatment	2009	N/A	Yes	No	N/A
Klang Aow 1 shipwreck	Middle of Gulf of Thailand, 89 km from Satta-hip, Chonburi	1500–1530	N/A	Low	Nontreatment	1991	N/A	Yes	No	Oceangoing trader
Ko Kradat shipwreck	1 km north of the northern end of the island in Trad	1522–66	Southeast Asia	No hull remains	Nontreatment	1977	N/A	Yes	Yes	Oceangoing trader
Ko Rin wreck site	N/A	1558–1757	N/A	No hull remains	N/A	1988	N/A	Yes	No	N/A
Ko Samae San shipwreck	N/A	1658–1857	N/A	N/A	N/A	1985	N/A	Yes	No	N/A
Ko Samui shipwreck	In the waters between Samui Island and Tan Island	1658–1857	N/A	Low	Nontreatment	1984	N/A	Yes	Yes	Oceangoing trader
Prachuap Khiri Khan wreck site	N/A	1558–1757	N/A	Low	N/A	1987	N/A	Yes	No	N/A
Pattaya shipwreck	In the waters between Pat-taya Beach and Lan Island	1558–1657	N/A	Low	Nontreatment	1977	N/A	Yes	Yes	Oceangoing trader
Rayong wreck site	N/A	1657–1757	N/A	N/A	N/A	1977	N/A	Yes	No	N/A

Bangkachai I	Bangkachai Bay	N/A	N/A	No hull remains	N/A	1989	N/A	Yes	No	N/A
Bangkachai II	In the waters 2 km from shore, Bangkachai Bay	early 17th century	Siamese?	Low	Nontreatment	1992	N/A	Yes	Yes	Oceangoing trader
Samed Ngam	Riverbank near mouth of Hanthabiuri River, Tambon Samedngam Aumper Muang, Chanthaburi	18th–19th century	Fujian junk?	Low	Preserved in situ in fresh water at local site museum	1982	N/A	Yes	Yes	Oceangoing trader
Indonesia										
Java Sea shipwreck	N/A	13th century?	Indonesian?	Low	Nontreatment	N/A	Yes	Yes	Yes	Indonesian trader
Pulau Buaya wreck	Pulau Buaya–Kepulauan Riau	1300s	Song dynasty	No hull remains	Other	1989	Yes	Yes	No	Trader
Belanakan shipwreck	Belanakan waters, Subang, West Java	14th century	N/A	N/A	N/A	N/A	N/A	N/A	N/A	N/A
Bukit Jakes shipwreck	Bintan Island	1400–1460	Ming dynasty	High	In situ preservation	1981?	No	Yes	Yes	Trader
Teluk Sumpat wreck	Bintan Island	1700s	Yuan and Ming dynasties	No hull remains	N/A	2005	Yes	Yes	Yes	N/A

Glossary

This brief glossary of nautical terms is intended to clarify medieval East Asian shipbuilding technology. The definitions of these terms rely by default on the Western nautical terms presented in the *International Maritime Dictionary* (Kerchove 1961) and follow terminology developed in previous studies of Chinese sailing ships by George Worcester (1971) and Hans Van Tilburg (2007). A few new terms resulting from the present research appear in this glossary to explain features of excavated ships in the region.

athwartships—At right angles to the fore-and-aft line of a vessel (Kerchove 1961).

balanced rudder—A rudder in which part (generally 12–20 percent) of the blade surface is forward of the axis so that the water pressure on this portion counterbalances that on the after part (Kerchove 1961).

bao shou kong—A *bao shou kong* (保寿孔) is literally a "hole for holding longevity," presumed to mean for safe journeying. The term applies to holes in the scarfs or joints of the keel of traditional Chinese ships, used for placing copper coins representing the Great Bear constellation and a mirror representing a full moon, both serving as navigational aids. This appears to have been a common tradition among shipwrights in the southern part of Fujian.

batten—A small spar, usually of bamboo, secured to and lying across a sail to extend the leech, or aft edge of the sail (Worcester 1966).

beam—A transverse structural member of the ship's framing. Beams serve as joists supporting the deck against pressures that may be sustained from cargo or from the masses of ties and struts they support; beams hold the vessel's sides at fixed distance and check racking tendencies in the transverse section (Kerchove 1961).

bow transom—The foremost transverse baulks forming the bow.

bracket—A fastening to fix the hull planking and bulkheads tightly together and to hold bulkhead planks.

bulkhead—A vertical partition, whether fore and aft or athwartships, that separates different compartments or spaces from one another (Kerchove 1961).

bulkhead stiffeners—Bars having either rolled section, angles, bulb angles, or channels, and welded or riveted vertically on one side of a bulkhead to strengthen it. The scantlings of stiffeners are regulated according to the water pressure to which the bulkhead may be subjected. Stiffeners are fitted with brackets or lug attachments at both ends (Kerchove 1961).

bulwark planks—Planks running along each side of the vessel, serving as a fence, to keep the decks dry (Kerchove 1961).

bulwark stanchion—In wooden hulls, the upper part of top timbers that extends above the sheer line and supporting the bulwark planks (Kerchove 1961).

butt—The ends of two planks that exactly meet endwise.

butt joint—The point of connection between two parts where their ends or edges are brought together and fastened with one of several joint methods, including lap joints and tongue and groove joints.

butt strap (plate)—A strip of plating that connects the two parts meeting at a butt joint.

carvel-built—Said of a boat constructed with edge-to-edge-fastened carvel planking (Kerchove 1961).

carvel planking—A system of planking in which the outside planking of a vessel or boat is flush: the edges meeting and giving the shell a smooth surface instead of the planks overlapping, as in the clinker system. The lines for carvel-built boats are laid down on a floor, just as in the building of large vessels; molds are made, and the timbers are beveled (Kerchove 1961).

chine—The line of intersection between the sides and the bottom of a flat or V-shaped hull (Kerchove 1961).

chine strake—The strakes create the chine between the bottom and the sides, specifically, for those planks appearing in the Goryeo dynasty ships having an L-shaped cross section.

chunam—A substance traditionally used in Chinese shipbuilding as putty for caulking and luting, generally a material made from calcareous substances (such as the ash of burned limestone or shells) and vegetable fiber mixed with wood oil (tung oil).

clinker-built—Said of a boat constructed with overlapping clinker planking (Kerchove 1961).

clinker planking—A method of planking used for watercraft in which the lower edge of each strake overlaps the upper edge of the next strake below (Kerchove 1961).

coaming—The raised borders around the edges of hatches and scuttles for preventing water on deck from running below (Kerchove 1961).

compound anchor—A type of anchor consisting of wooden shanks and arms with a stone anchor stock (stone stocks).

deadrise—An angle measured upward from a horizontal plane at the lowest point of the keel (Kerchove 1961).

deadweight—The vessel's carrying capacity when loaded: includes crew, passengers, ballast, fresh water in storage tanks, and cargo (Kerchove 1961).

displacement—The number of tons of water displaced by a vessel afloat (Kerchove 1961).

draft—The depth of water a ship requires to float freely, or the depth of a vessel below the waterline, measured vertically to the lowest part of the hull or other reference points (Kerchove 1961).

dunnage—Material stowed among and beneath the cargo of a vessel to protect the cargo from chafing and getting wet (Kerchove 1961).

frame—A transverse structural member forming the ribs of the hull and extending from the keel to the highest continuous deck (Kerchove 1961). For oceangoing ships of East Asian origin, half frames attached to the bulkheads are used.

frame-fast technique—The procedure for constructing the hull by first erecting frames and then attaching hull planks to them.

freeboard—The vertical distance from the loaded waterline to the uppermost complete deck (Kerchove 1961).

gua ju—A *gua ju* (挂锔) is literally a "hook bracket." The use of metal brackets in the Chinese shipbuilding tradition was introduced following examples of wooden brackets and various iron cramps (Xu 1985; Zhang et al. 2004: 288–290). The procedure of installation may show similarities with an L-shaped wrought iron spike or dog as used in recent times to fasten the bulkheads to hull bottom planks of Chinese river junks. Joseph Needham (1971: 115) mentions the size of metal brackets. Worcester's work provides details on brackets used in traditional Chinese shipbuilding (Worcester 1941, 1971).

half-breadth plan—A plan or top view of half of a ship, divided longitudinally (Kerchove 1961).

half frame—The short frame timbers or, for medieval oceangoing ships of East Asian origin, a frame consisting of two members or timbers meeting in the center line of the ship.

hog piece—A fore-and-aft piece of timber fastened to the top of the keel, providing a good landing edge for the garboards (Kerchove 1961). In some traditional Chinese ships, it is used as a solid longitudinal foundation where transverse components such as bulkheads can sit neatly.

junk—A term used to describe a wooden sailing vessel of Oriental build and rig or, more specifically, used in Western nautical terminology to describe a traditional type of Chinese ship. The term has been used only by foreigners to describe watercraft operating in Chinese waters; there is no Chinese ideogram with the same meaning, apart from a substitute character (戎克) tentatively used by the Japanese.

keel—Known as *longwu* (竜骨) in Chinese shipbuilding terminology, a keel is likely to consist of three timbers: a main keel, a forward keel, and an aft keel. The center keel is definitive as a main keel in that it underlies the main internal structure, namely, the bulkheads. Bulkheads are also built over the aft keel. The forward keel functions as an extension of the main keel but is not likely to underlie bulkheads. The term *longwu* is therefore applicable only to the main and aft keel members. As ideogrammatically rendered in Chinese characters, it means the backbone of the ship, physically supporting the upper parts of the hull. Most of the archaeologically identified ships originating from eastern Asia show a similar longitudinal keel structure based on three components.

leeboard—A plate or frame of planks lowered over the lee side of a shallow-draft boat with a flat bottom. It lessens the leeway by giving increased lateral resistance (Kerchove 1961).

light displacement—The weight of a vessel when unloaded (also called light weight), including the hull, fittings, permanent ballast, and feed water (Kerchove 1961).

limber holes—Holes cut for drainage purposes in the lower part of frames or bulkheads in the center line of the ship or in the vicinity of the keel.

mast heel—The lower end of a mast.

mast step—The chamfered block where the heel of a mast meets the tabernacle cheeks flanking it likely to have two square recesses to accommodate the ends of the tabernacle cheeks.

planked-up dugout—A type of dugout canoe made by extending its sides upward by adding extra side strakes.

plank-fast technique—The procedure for constructing the hull that proceeds by first building up a plank shell, in comparison with the frame-fast technique.

protruding-tongue scarf—A scarf joint showing a half lap with a protruding tongue, used in the bottom strakes of Goryeo dynasty ships. A protruding tongue like a tenon is made in the joint of a longitudinal member, which joins into the recess cut in another member.

rabbeted carvel seam—A seam in which "the edge-joint between two adjacent strakes is rabbeted along the whole of the seam by a type of step-joint"

(Burningham and Green 1997: 36) The term was developed for the Quanzhou ship's hull (Burningham and Green 1997).

rabbeted clinker seam—A seam in which "a rabbet is cut into the inside of the lower edge of the upper strake; the upper (unrabbeted) edge of the lower [strake] is set in this rabbet, giving an external appearance of a clinker overlap, but the thickness of the step between the strakes at the surface is reduced by the depth of [the] rabbet" (Burningham and Green 1997: 37). In the case of the Shinan ship, rabbets were cut in both the lower and the upper edges of some strakes around the bow where the hull planking changes from clinker- to carvel-built.

rudder blade—The main flat portion of the rudder that provides the necessary surface for the impinging action and side pressure of the water (Kerchove 1961).

rudder gudgeon (rudder socket)—A device retaining a rudder. Three different types of rudder gudgeons (sockets) have been identified, according to Worcester's study: open, half-open, and closed gudgeons (Worcester 1971).

rudder post—A part of a rudder that acts as a shaft to receive a rudder blade.

rudder tiller—A heavy bar or lever with one end of the board fitting onto the rudder post. Its function is to turn the rudder (Kerchove 1961).

sacrificial planking—A covering of wooden planks fastened to the underwater body of a wooden hull to protect against shipworms and marine growths of all kinds.

scarf—A join whereby two members are connected longitudinally into a continuous piece, the ends being halved, notched, or cut away so as to fit into each other with mutual overlapping. Typical scarf conformations in East Asian shipbuilding traditions are a stopped scarf and a hooked scarf.

scupper—One of the drains set in a deck to carry off the accumulation of rain- or seawater (Kerchove 1961).

scupper hole—A drain hole cut through the gunwale angle bar and adjoining shell plate to allow water to run directly overboard from the gutter or waterway (Kerchove 1961).

sheer plan—A plan showing the longitudinal, vertical section passing through the median line of a ship (Kerchove 1961).

skewed nail—A type of iron nail, its shaft bent, typically used for hull planking. In Chinese shipbuilding, various types of iron nails were used for planking, including *diao ding* (literally "hanging nails," 吊钉), *bie ding* (literally "split nails," 别钉), and *cha ding* (literally "spike nails," 插钉), and each is distinctive in their shape, length, and cross section (Li 1989).

stern transom—Aftermost athwartship baulks giving shape to and forming a stern, partly supporting the extremities of the deck. See also *transom stern*.

stiffeners—Also called stiffening bars. Sections or square shapes used for increasing the rigidity of plating, as in bulkhead stiffeners (Kerchove 1961). See also *bulkhead stiffeners*.

strake—A continuous line of planking in the hull extending the whole length of the vessel. There are several varieties of strakes, according to their position; e.g., planks at the bottom of Korean ships are described as bottom strakes.

structural bulkhead—A bulkhead, generally water- or oil-tight, that forms one of the boundaries of the main compartments and contributes to the strength of the hull. Its depth usually extends through several decks (Kerchove 1961).

tabernacle—A vertical trunk to support the lower part of the mast.

tabernacle cheeks—Vertical partners on each side of the lower end of the mast.

transom-mounted rudder—A rudder inserted into a round aperture cut through the stern transom or a hole in the gudgeons fixed onto the stern transom. See also *rudder gudgeons*.

transom stern—A form of stern in which the upper part of the aft hull structure terminates in a large flat surface, vertical, or nearly so, and square to the central longitudinal plane of the ship.

turret construction—A barrel-shaped hull with a comparatively narrow deck superimposed on it (Worcester 1966).

waterline—Horizontal lines parallel to the base line in the body plan or the sheer plan or horizontal lines parallel to curved lines in the half-breadth plan (Kerchove 1961).

watertight bulkheads—Vertical partitions in a ship that are strongly built, with joints tight enough to withstand the pressure of the hydrostatic head and prevent water from escaping into adjoining spaces if a compartment fills with water. Watertight bulkheads also contribute to the structural strength of the ship (Kerchove 1961).

windlass—A type of winch used to hoist the anchors, house them safely, and warp or kedge the ship when in harbor (Kerchove 1961).

yuloh—A type of sculling oar used in small Chinese watercraft, placed in the stern and consisting of two parts, a shaft and a blade (Kerchove 1961).

References

Abe, Y. 2010. Tojiki kara mita hokubu Betonamu no boeki yoso: Van Don to Fohien no kokogaku [Trades in northern Vietnam based on ceramic study: Archaeology on Van Don and Pho Hien]. In *Umi no michi to kokogaku: Indoshina Hanto kara Nihon* [Sea route and archaeology: From the Indochinese peninsula to Japan], ed. Seiichi Kikuchi and Yuriko Abe, 127–38. Tokyo: Koshi Shoin.

Adachi, H. 1998. *Nihon no fune: Wasen hen* [Ships in Japan: Traditional Japanese ships]. Tokyo: Museum of Maritime Science.

Adams, Jeff. 2013. The role of underwater archaeology in framing and facilitating the Chinese national strategic agenda. In *Cultural heritage politics in China*, ed. Tami Blumenfield and Helaine Silverman, 261–82. New York: Springer.

Adams, Jonathan. 2001. Ships and boats as archaeological source material. *World Archaeology* 32 (3): 292–10.

Agency for Cultural Affairs. 2000. *Iseki hozon hoho no kento: Suichuu iseki* [Considering methods for protection of archaeological site: Underwater site]. Tokyo: Agency for Cultural Affairs.

Atkinson, K. 1990. The first stage of a maritime archaeological training programme in the People's Republic of China: A brief summary of the aims, procedures and results. *International Journal of Nautical Archaeology* 19 (3): 255–58.

Atkinson, K., J. Green, R. Harper, and V. Intakosai. 1989. Joint Thai-Australian underwater archaeological project, 1987–1988: Part 1, Archaeological survey of wreck sites in the Gulf of Thailand, 1987–1988. *International Journal of Nautical Archaeology* 18 (4): 299–315.

Atmajuana, W., and E. E. McKinnon. 2005. The Jepara wreck: C. 12th century ceramics from Fujian and Zhejiang. In *Proceedings of international conference: Chinese Export Ceramics and Maritime Trade, 12th–15th Century*, ed. Pei-kai Cheng, Guo Li, and C. K. Wan, 126–42. Hong Kong: Chung Hwa Book Co.

Basch, L. 1972. Ancient wrecks and the archaeology of ships. *International Journal of Nautical Archaeology* 1: 1–58.

Beavan, N., T. Sokha, U. Zoppi, B. McCarthy, M. Schilling, L. Cort, and S. F. Lu. 2012. Field note: A radiocarbon date for the Koh S'dech shipwreck, Koh Kong Province, Kingdom of Cambodia. Freer Sackler Gallery, SEA Ceramics Library, Smithsonian Institute. http://seasianceramics.asia.si.edu/resources/essay.asp?id=218.

Bellwood, P. 1985. *Prehistory of the Indo-Malaysian Archipelago*. Sydney: Academic Press.

Bellwood, P., J. Cameron, N. V. Viet, and B. V. Liem. 2007. Ancient boats, boat timbers, and locked mortise-and-tenon joints from Bronze/Iron Age northern Vietnam. *International Journal of Nautical Archaeology* 36 (1): 2–20.

Beltrame, C., and D. Gaddi. 2007. Preliminary analysis of the hull of the Roman ship from Grado, Gorizia, Italy. *International Journal of Nautical Archaeology* 36 (1): 138–47.

Bernard, W. D. 1844. *Narrative of the voyages and services of the* Nemesis *from 1840 to 1843, and of the combined naval and military operations in China, comprising a complete account of the colony of Hong-Kong and remarks on the character and habits of the Chinese / from notes of Commander W. H. Hall, R. N.; with personal observations by W. D. Bernard*. 2nd ed. London: Henry Colburn.

Biles, J. H. 1908. *The design and construction of ships*. London: C. Griffin.

Blot, J. Y. 2006. Naval engineering concepts and cultural bias in the analysis of longitudinal stiffness when applied to ancient Chinese shipbuilding. Paper presented at International Symposium on Penglai Ancient Ship, Penglai.

Burningham, N. 1989. The structure of Javanese Perahu. *The Beagle: Records of the Northern Territory Museum of Arts and Sciences* 6 (1): 195–219.

Burningham, N., and J. Green. 1997. Description of the Quanzhou ship. In *Maritime archaeology in the People's Republic of China*, ed. Jeremy Green, 32–48. Special Publication no. 1, Australian National Centre of Excellence for Maritime Archaeology. Fremantle: Western Australian Museum.

Cai, W., C. Li, and L. Xi. 2010. Watertight bulkheads and limber holes in ancient Chinese boats. In *Shipwreck ASIA: Thematic studies in East Asian maritime archaeology*, ed. Jun Kimura, 26–31. Adelaide: Maritime Archaeology Program.

Caira, G. 2008. The Cholne: An Iron Age dugout canoe from Russia. *Newsletter of the Nautical Archaeology Society* 1: 14–15.

Cha, M. Y. 2011. Micromorphological and chemical characteristics of waterlogged archaeological bamboo. In *Asia-Pacific Regional Conference on Underwater Cultural Heritage Proceedings*, ed. Mark Staniforth, Jennifer Craig, Sheldon Clyde Jago-on, Bobby Orillaneda, and Ligaya S. P. Lacsina. The MUA Collection. Manila: National Museum of the Philippines. www.themua.org/collections/files/original/7e7546f422b66ecda94d7ad145bd1d4b.pdf.

Chaffee, J. W. 2001. The impact of the Song: Imperial clan on the overseas trade of Quanzhou. In *The emporium of the world: Maritime Quanzhou, 1000–1400*, ed. Angela Schottenhammer, 13–46. Leyden: Brill.

———. 2008. Muslim merchants and Quanzhou in the late Yuan–early Ming: Conjectures on the ending of the medieval Muslim trade diaspora. In *The East Asian "Mediterranean": Maritime crossroads of culture, commerce and human migration*, ed. Angela Schottenhammer, 115–32. Wiesbaden: Harrasoowitz Verlag.

———. 2010. Song China and the multi-state and commercial world of East Asia. *Crossroads: Studies on the History of Exchange Relations in the East Asian World* 1–2: 33–54.

Chakravarti, R. 2002. Seafarings, ships and ship owners: India and the Indian Ocean

(AD 700–1500). In *Ships and the development of maritime technology in the Indian Ocean*, ed. David Parkin and Ruth Barnes, 28–61. London: Routledge Curzon.

Chang, T.-t. 1934. *Sino-Portuguese trade from 1514 to 1644: A synthesis of Portuguese and Chinese sources*. Leyden: E. J. Brill.

Cheng, J. (ed.). 1985. *Mu cai xue* [Timber study]. Beijing: Zhongguo Linye Chubanshe.

Choi, H.-S. 2004. A naval architectural study on the sunken ship at Shinan, Korea. In *The conservation and restoration of Shinan ship: The 20 years history*, ed. National Maritime Museum of Korea, 162–69. Mokpo.

Choi, W.-G. 2006. *The traditional ships of Korea*. Trans. Jean Young Lee. Seoul: Ewha Womans University Press.

Church, S. K. 2005. Zheng He: An investigation into the plausibility of 450-ft treasure ships. *Monumenta Serica* 53: 1–42.

———. 2008. The Ming Treasure Shipyard archaeological report. In *ACUA underwater archaeology proceedings 2008*, ed. Susan Langley and Victor Mastone. Advisory Council on Underwater Archaeology. 30–38. Columbus, Ohio: PAST Foundation.

———. 2010. Two Ming dynasty shipyards in Nanjing and their infrastructure. In *Shipwreck ASIA: Thematic studies in East Asian maritime archaeology*, ed. Jun Kimura, 32–49. Adelaide: Maritime Archaeology Program.

Clark, H. R. 1991. *Community, trade, and networks: Southern Fujian Province from the third to the thirteenth century*. Cambridge, U.K.: Cambridge University Press.

Clark, P., E. Conese, N. Nicolas, and J. Green. 1989. Philippines archaeological site survey, February 1988. *International Journal of Nautical Archaeology* 18 (3): 255–62.

Clark, P., J. Green, T. Vosmer, and R. Santiago. 1993. The Butuan Two boat known as a *balangay* in the National Museum, Manila, Philippines. *International Journal of Nautical Archaeology* 22 (2): 143–59.

Cui, Y. 2009. Comparative study of traditional boat construction techniques in Korea and Japan. *Maritime History Studies*, no. 1: 45–54.

Cui, Y., and S.-j. Kim. 2007. Significance of Penglai Koryo ancient ship in the history of Korean shipbuilding. *Maritime History Studies*, no. 2: 99–108.

Cultural Relics and Archaeological Institute of Shandong Province, Yantai Municipal Museum, and Cultural Relics Bureau of Penglai City. 2006. *Ancient ships from Penglai*. Beijing: Wenwu Press.

Dai, K. 1985. Notes on the origination of ancient Chinese junks based upon study of unearthed dug-out canoes. *Marine History Research, Transactions of CSNAME Marine History Research Group* 1: 4–17.

Danjo, H. 2005. Min dai kaikin no jitsuzo [Fact of sea ban in Ming dynasty]. In *Minatomachi to kaiiki sekai* [Port city and the maritime sphere], ed. Shosuke Murai, 145–77. Minatomachi no sekaishi [Series world history of the port cities]. Tokyo: Aoki Shoten.

Deguchi, A. 1995. *Nippon to syuhen Ajia no dentou-teki-senpaku* [Traditional ships in Japan and the Asian region: The study of cultural geography]. Tokyo: Bunken Shuppan.

Delgado, J. P. 2009. *Khubilai Khan's lost fleet: In search of a legendary armada*. Berkeley: University of California Press.

Dizon, E. Z. 1992. Report on underwater archaeological activities in the Philippines from 1991 to mid-1992. In *Final report of the SEAMEO-SPAFA consultative workshop on underwater archaeological research*, 39–49. Jakarta: SEAMEO-SPAFA.

Donnelly, I. A. 1923. Foochow pole junks. *Mariner's Mirror* 9: 226–31.

———. 1924. River craft of the Yangtzekiang. *Mariner's Mirror* 10: 4–16.

———. 1925. Early Chinese ships and trade. *Mariner's Mirror* 11: 344–54.

———. 1926. The "oculus" in Chinese craft. *Mariner's Mirror* 12: 339–40.

———. 1930. *Chinese junks and other native craft*. 2nd ed. Shanghai: Kelly and Walsh.

———. 1936. Strange craft of China's inland waters. *Mariner's Mirror* 22: 410–21.

Doran, E. 1967. The origin of leeboards. *Mariner's Mirror* 53: 39–57.

Duivenvoorde, W. V. 2014. The 5th-century BC shipwreck at Tektaş Burnu, Turkey: Evidence for the ship's hull from nail concretions. *International Journal of Nautical Archaeology* 43 (1): 10–26.

Enomoto, W. 2008. Chugoku jin no kaijo sinsyutsu to kaijo teikoku toshiteno Chugoku [Advent of the Chinese merchant in maritime activities and China as a maritime empire]. In *Kaiiki Asia shi kenkyu nyumon* [A research guide to maritime Asian history], ed. Shiro Momoki, 16–22. Tokyo: Iwanami Shoten.

Farris, W. W. 2009. Shipbuilding and nautical technology in Japanese maritime history: Origins to 1600. *Mariner's Mirror* 95 (3): 260–83.

Flatman, J., and M. Staniforth. 2006. Historical maritime archaeology. In *The Cambridge companion to historical archaeology*, ed. Dan Hicks and Mary C. Beaudry, 168–88. Cambridge, U.K.: Cambridge University Press.

Flecker, M. 2002. *The archaeological excavation of the 10th century Intan shipwreck*. British Archaeological Reports S1047. Oxford, U.K.: Archaeopress.

———. 2007. The South China Sea tradition: The hybrid hulls of South-East Asia. *International Journal of Nautical Archaeology* 36 (1): 75–90.

———. 2012. The Jade Dragon wreck: Sabah, East Malaysia. *Mariner's Mirror* 98 (1): 9–29.

Gibbins, D., and J. Adams. 2001. Shipwrecks and maritime archaeology. *World Archaeology* 32 (3): 279–91.

Glover, I., and P. Bellwood (eds.). 2004. *Southeast Asia: From prehistory to history*. London: Routledge Curzon.

Goddio, F., M. Crick, P. Lam, S. Pierson, and R. Scott. 2002. *Lost at sea: The strange route of the Lena Shoal junk*. London: Periplus Publishing.

Gong, C., Y. Ding, X. Chu, and L. Xi. 2008. Zhejiang Ningbo Heyilu chutu guchuan fuyuan yanjiu [Study on the reconstruction of the ancient ship from Zhejiang Ningbo Heyilu]. *Ningbo Wenwu Kaogu Yonjiu Wenshi* [Bulletin for research of cultural relics and archaeology in Ningbo] 2: 183–94.

Gong, C., and L. Xi. 2006. The exchange of navigation technology between China and Korea: The case of two ancient Korean boats unearthed in Penglai, Shandong. *Maritime History Studies*, no. 2: 20–22.

Gould, R. 2000. *Archaeology and the social history of ships*. Cambridge, U.K.: Cambridge University Press.

Green, J. 1983a. The Song dynasty shipwreck at Quanzhou, Fujian Province, People's Republic of China. *International Journal of Nautical Archaeology* 12 (3): 253–61.

———. 1983b. The Shinan excavation, Korea: An interim report on the hull structure. *International Journal of Nautical Archaeology* 12 (4): 293–301.

———. 1986. Eastern shipbuilding traditions: A review of the evidence. *Bulletin of the Australasian Institute for Maritime Archaeology* 10 (2): 1–6.

———. 1990. Maritime archaeology in Southeast and East Asia. *Antiquity* 64: 347–63.

———. 1997a. Archaeological evidence East Asian vessels. In *Maritime archaeology in the People's Republic of China*, ed. Jeremy Green, 19–24. Special Publication no. 1, Australian National Centre of Excellence for Maritime Archaeology. Fremantle: Western Australian Museum.

———. 1997b. Chinese shipbuilding in a historical context. In *Maritime archaeology in the People's Republic of China*, ed. Jeremy Green, 1–18. Special Publication no. 1, Australian National Centre of Excellence for Maritime Archaeology. Fremantle: Western Australian Museum.

———. 1998. The archaeological contribute to the knowledge of the extra-European shipbuilding at the time of the medieval and modern Iberian-Atlantic tradition. Paper presented at the International Symposium on Archaeology of Medieval and Modern Ships of Iberian-Atlantic Tradition, Lisbon.

———. 2006. Nautical archaeology in Australia, the Indian Ocean and Asia. In *Maritime archaeology: Australian approaches*, ed. Mark Staniforth and Michael Nash, 97–109. New York: Springer.

Green, J., N. Burningham, and Museum of Overseas Communication History. 1998. The ship from Quanzhou, Fujian Province, People's Republic of China. *International Journal of Nautical Archaeology* 27 (4): 277–301.

Green, J., and P. Clark. 1993. *The maritime archaeological, historical and ethnographic investigation of maritime technology in the pre-modern period, [with] particular reference to Asian built ships*. Western Australian Maritime Museum Department of Maritime Archaeology Report 68. Fremantle.

Green, J., P. Clark, and K. Millar. 1993. *The academies' Australia-China exchange in the humanities and social science: Report on a visit to China*. Western Australian Maritime Museum Department of Maritime Archaeology Report 67. Fremantle.

Green, J., and R. Harper. 1982. The excavation of the Ko Kradat wreck site, Thailand. *International Journal of Nautical Archaeology* 11 (2): 164–71.

———. 1983. *The excavation of the Pattaya wreck site and survey of three other sites, Thailand, 1982*. Special Publication no. 1. Fremantle: Australian Institution for Maritime Archaeology.

Green, J., R. Harper, and V. Intakosi. 1986. The Ko Si Chang One shipwreck excavation, 1983–1985: A progress report. *International Journal of Nautical Archaeology and Underwater Exploration* 15 (2): 105–22.

———. 1987a. *The maritime archaeology of shipwrecks and ceramics in Southeast Asia: The maritime connection*. Albert Park: Australian Institute for Maritime Archaeology.

———. 1987b. *The Ko Si Chang Three shipwreck excavation, 1986*. Special Publication no. 4. Albert Park: Australian Institute for Maritime Archaeology.

Green, J., and V. Intakosai. 1983. The Pattaya wreck site excavation, Thailand: An interim report. *International Journal of Nautical Archaeology and Underwater Exploration* 12 (1): 3–13.

Green, J., and Z. G. Kim. 1989. The Shinan and Wando sites, Korea: Further information. *International Journal of Nautical Archaeology* 18 (2): 33–41.

Green, J., T. Vosmer, P. Clark, R. Santiago, and M. Alvares. 1992. *Interim report on the joint Australian-Philippines Butuan boat project, October 1992*. Fremantle: Western Australian Maritime Museum.

———. 1995. Interim report on the joint Australian-Philippines Butuan boat project, October 1992. *International Journal of Nautical Archaeology* 24 (3): 177–88.

Greenhill, B. 1976. *Archaeology of the boat: A new introductory study*. London: A. and C. Black.

Guangdongsheng Wenwu Kaogu Yanjiusuo (ed.). 2014. *Gufan yizhen: Nanao 1 hao chushui wenwu jingpin tulu* [Shipwreck site: Selected artifacts from the Nanao no. 1]. Beijing: Science Press.

Guangzhou Shi Wenhuaju. 1999. *Guangzhou Qin Han kao gu san da fa xian* [Three archaeological finds of the Qin and Han periods in Guangzhou]. Guangzhou: Guangzhou Chu Ban She.

Guy, J. 1980. *Oriental trade ceramics in Southeast Asia, 10th to 16th century: Selected from Australian collections, including the Art Gallery of South Australia and the Bodor Collection*. Melbourne: National Gallery of Victoria.

———. 2004. The Intan shipwreck: A 10th-century cargo in South-East Asian waters. In *Song ceramics: Art history, archaeology and technology*, ed. Stacey Pierson, 171–92. Percival David Foundation of Chinese Art, School of Oriental and African Studies. London: University of London.

———. 2006. Early ninth-century Chinese export ceramics and the Persian Gulf connection: The Belitung shipwreck evidence. *SFECO (Société française d'études de la céramique orientale)* 4: 9–20.

Hall, K. R. 1985. *Maritime trade and state development in early Southeast Asia*. Honolulu: University of Hawaii Press.

Harrison, T. 1976. Brunei's two (or more) capitals and other ancient centres. *Brunei Museum Journal* 3 (4): 77–86.

Hartwell, R. 1962. A revolution in the Chinese iron and coal industries during the Northern Sung, 960–1126 A.D. *Journal of Asian Studies* 21 (2): 153–62.

———. 1966. Markets, technology, and the structure of enterprise in the development of the eleventh-century Chinese iron and steel industry. *Journal of Economic History* 26 (1): 29–58.

Hayashida, K. 2006. Present and future of underwater archaeology in Japan. In *Shinan Underwater Relics and 14th Century Asian Marine Trades: International symposium in celebration of the 30th anniversary of the Shinan wreck excavation*, ed. National Maritime Museum of Korea, 435–46. Mokpo.

Heng, D. 2009. *Sino-Malay trade and diplomacy from the tenth through the fourteenth century*. Athens: Ohio University Press.

Higham, C. 1989. *The archaeology of mainland Southeast Asia: From 10,000 B.C. to the fall of Angkor*. Cambridge, U.K.: Cambridge University Press.

Hoagland, P. 1997. Managing the underwater cultural resources of the China seas: A comparison of public policies in mainland China and Taiwan. *International Journal of Marine and Coastal Law* 12 (2): 265–83.

———. 1999. Historic shipwreck policy in China. In *Legal protection of the underwater cultural heritage: National and international perspectives*, ed. Sarah Dromgoole, 19–41. London: Kluwer Law International.

Hocker, F. M. 2004. Shipbuilding: Philosophy, practice, and research. In *The philosophy of shipbuilding: Conceptual approaches to the study of wooden ships*, ed. Frederick M. Hocker and Cheryl A. Bass, 1–11. College Station: Texas A&M University Press.

Hornell, J. 1934. The origin of the junk and sampan. *Mariner's Mirror* 30: 331–37.

Horridge, G. A. 1982. *The lashed-lug boat of the eastern archipelagoes, the Alcina MS and the Lomblem whaling boats*. Maritime Monographs and Reports. London: Trustees of the National Maritime Museum.

Hourani, G. F. 1963. *Arab seafaring in the Indian Ocean in ancient and early medieval times*. Beirut: Khayats.

Huisman, H.D.J. 2009. Iron. In *Degradation of archaeological remains*, ed. Hans D. J. Huisman, 91–109. The Hague: Sdu Uitgevers.

Intakosai, V. 1984. Country report of Thailand. In *SPAFA consultative workshop on research on maritime shipping and trade networks in Southeast Asia*, 133–43. Cisarusa: SEAMEO-SPAFA.

Ishii, K. 1957. *Nihon no fune* [Ships in Japan]. Tokyo: Sogensha.

———. 1995a. *Wasen I* [A Japanese vessel I]. Mono to ningen no bunkashi [Cultural history of materials and humans]. 3rd ed. Tokyo: Hosei Daigaku Syuppankyoku (Hosei University Press).

———. 1995b. *Wasen II* [A Japanese vessel II]. Mono to ningen no bunkashi [Cultural history of materials and humans]. 3rd ed. Tokyo: Hosei Daigaku Syuppankyoku (Hosei University Press).

Iwao, S. 1958. *Shuinsen boekishi no kenkyu* [Research on the history of the vermilion-seal ship]. Tokyo: Kobundo.

Ji, S. 1987. Brief report on the ancient boat from the Dazhi River, Nahui country, Shanghai. *Bulletin of the Shanghai Museum* 4: 175–78.

Jian, S. 2011. Remaining treasures of the Maritime Silk Road: The excavation of Ming dynasty shipwreck no. 1, Nan'ao. In *2010 International Meeting on Protection, Presentation and Valorisation of Underwater Cultural Heritage*, ed. Shuguang Liu, 291–97. Beijing: Wenwu Press.

Johnson, M. 1999. *Archaeological theory: An introduction*. Oxford, U.K.: Blackwell.

Kabanov, Y., and P. Pomey. 2004. The Greek sewn shipbuilding tradition and the *ma'agan mikhael* ship: A comparison with Mediterranean parallels from the sixth to the fourth centuries BC. *Mariner's Mirror* 90 (1): 6–28.

Kamei, M. 1993. Trade routes of trade ceramics in the Nansei Islands, Japan. *Journal of Southeast Asian Studies* 11: 11–45.

Kano, T., and K. Segawa. 1956. *An illustrated ethnography of Formosan aborigines*. Rev. ed. Tokyo: Maruzen.

Katagiri, C. (ed.). 2007. *Engan chiiki iseki bunpu chosa gaiho I: Okinawa Honto oyobi shuuhen rito hen* [Brief report of offshore sites, part 1, Okinawa main island and surrounding islands]. Nakagami: Archaeological Center of Okinawa.

——— (ed.). 2009. *Engan chiiki iseki bunpu chosa gaiho II: Miyako-Yaeyama Shoto hen* [Brief report of offshore sites, part 2, Miyako-Yaeyama Islands]. Nakagami: Archaeological Center of Okinawa.

Keith, D. H. 1980. A fourteenth-century shipwreck at Sinan-gun. *Archaeology* 33 (2): 33–43.

Keith, D. H., and C. J. Buys. 1981. New light on medieval Chinese seagoing ship construction. *International Journal of Nautical Archaeology* 10 (20): 119–32.

Kenderdine, S. 1995. Bai Jiao 1—The excavation of a Song dynasty shipwreck in the Dinghai area, Fujian Province, China, 1995. *International Journal of Nautical Archaeology* 24 (4): 247–66.

———. 1997. The Bai Jiao excavations. In *Maritime archaeology in the People's Republic of China*, ed. Jeremy Green, 53–93. Special Publication no. 1, Australian National Centre of Excellence for Maritime Archaeology. Fremantle: Western Australian Museum.

Kerchove, R. D. 1961. *International maritime dictionary: An encyclopedic dictionary of useful maritime terms and phrases, together with equivalents in French and German*. New York: D. Van Nostrand.

Kikuchi, S. 2010. Betonamu kaiiki no Chinbotsusen [Shipwrecks in Vietnamese waters]. In *Umi no michi to kokogaku: Indoshina Hanto kara Nihon* [Sea route and archaeology: From the Indochinese peninsula to Japan], ed. Seiichi Kikuchi and Yuriko Abe, 139–55. Tokyo: Koshi Shoin.

Kim, M., and W.-s. Moon. 2011. Tracking 800-year-old shipments: An archaeological investigation of the Mado shipwreck cargo, Taean, Korea. *Journal of Maritime Archaeology* 6: 129–49.

Kim, S.-B. 2006. Current situation and perspective of underwater archaeology in Korea and Shinan underwater relics. In *Shinan Underwater Relics and 14th Century Asian Marine Trades: International symposium in celebration of the 30th anniversary of the Shinan wreck excavation*, ed. National Maritime Museum of Korea, 389–421. Mokpo.

Kim, Y.-H. 2004. A study on the structure of Shinan shipwreck. In *The conservation and restoration of Shinan ship: The 20 years history*, ed. National Maritime Museum of Korea, 106–21. Mokpo.

Kim, Z.-G. 1991. The wreck excavated for the Wando-Island. In *International Sailing Ships History Conference*, ed. Shanghai Society of Naval Architecture and Marine Engineering and Marine History Researchers' Association of CSNAME, 56–58. Shanghai.

———. 1994. *Sok hanguk seonbaksa yeongu* [Research sequel to the history of Korean ships]. Seoul: Seoul National University Press.

Kimura, J. 2006. Recent survey and excavation on the Mongolian fleet sunk off Japan: The Takashima underwater site. *Bulletin of the Australasian Institute for Maritime Archaeology* 30: 7–13.

———. 2009. Kokunai suichuu iseki no hogo to kanri, bunkaisan toshiteno mondai [Issues of protection and management of domestic underwater sites as cultural heritage]. In *The collection of treaties for 30th anniversary of the Department of Archaeology*, ed. Department of Archaeology, Tokai University, 371–82. Tokyo: Rokuichi Shobo.

———. 2010a. Historical development of shipbuilding technologies in East Asia. In *Shipwreck ASIA: Thematic studies in East Asian maritime archaeology*, ed. Jun Kimura, 1–25. Adelaide: Maritime Archaeology Program.

——— (ed.). 2010b. *Shipwreck ASIA: Thematic studies in East Asian maritime archaeology*. Adelaide: Maritime Archaeology Program.

———. 2015. Shipwrecked? New developments in Southeast Asian maritime archaeology and the safeguard of shipwrecks in the region. *International Journal of Nautical Archaeology* 44 (1): 1–4.

Kimura, J., R. Sasaki, and V. T. Long. 2011. Historical development of Asian anchors, as evidenced by two wooden anchors found in northern Vietnam. *International Journal of Nautical Archaeology* 40 (1): 361–73.

Kimura, J., M. Staniforth, L. T. Lien, and R. Sasaki. 2014. Naval battlefield archaeology of the lost Kublai Khan fleets. *International Journal of Nautical Archaeology* 43 (1): 76–86.

Kinoshita, N. 2014. Relationship between Taiwan, the Ryukyus and Fujian during the 10th–15/16th centuries as seen from trade porcelain. *Kumamoto Journal of Culture and Humanities* 105: 35–51.

Kuwabara, J. 1989. *Ho Juko no jiseki* [Records of Pu Shougeng]. Toyo Bunko. Tokyo: Heibonsha.

Kuwata, R. 1954. Note on the ancient history of Southeast Asia. *Memoires of the Faculty of Literature, Osaka University* 3: 1–43.

Lacsina, L.S.P. 2011. Traditional island Southeast Asian watercraft in Philippine archaeological sites. In *Asia-Pacific Regional Conference on Underwater Cultural Heritage Proceedings*, ed. Mark Staniforth, Jennifer Craig, Sheldon Clyde Jago-on, Bobby Orillaneda, and Ligaya S. P. Lacsina. The MUA Collection. Manila: National Museum of the Philippines. www.themua.org/collections/files/original/3c6f3519be2b1385dfd92ecf2e54046c.pdf.

Laichen, S. 2010. Assessing the Ming role in China's southern expansion. In *Southeast Asia in the fifteenth century: The China factor*, ed. Geoff Wade and Sun Laichen, 44–79. Singapore: NUS Press.

Lape, P. V., S. O'Connor, and N. Burningham. 2007. Rock art: A potential source of information about past maritime technology in the South-East Asia–Pacific region. *International Journal of Nautical Archaeology* 36 (2): 238–53.

Lee, C.-E. 1991. A study on the structural and fluid characteristics of a rabbetted clinker type ship (the sunken ship salvaged off Shinan). In *International Sailing Ships History Conference*, ed. Shanghai Society of Naval Architecture and Marine Engineering and Marine History Researchers' Association of CSNAME, 154–68. Shanghai.

Lee, C.-H. 2004. The study on the restoration of the Shinan ship. In *The conservation and restoration of Shinan ship: The 20 years history*, ed. National Maritime Museum of Korea, 144–59. Mokpo.

L'Hour, M. (ed.). 2003. *The sunken treasures of Brunei Darussalam*. Sydney: Art Exhibitions Australia.

———. 2006. The Brunei shipwreck excavation: Asian maritime trade along the Borneo coast at the end of the 15th century. In *Shinan Underwater Relics and 14th Century*

Asian Marine Trades: International symposium in celebration of the 30th anniversary of the Shinan wreck excavation, ed. National Maritime Museum of Korea, 539–48. Mokpo.

Li, G. 1989. Archaeological evidence for the use of "chu-nam" on the 13th century Quanzhou ship, Fujian Province, China. *International Journal of Nautical Archaeology* 18 (4): 277–83.

———. 1998. Ancient Chinese anchors: Their rigging and conservation. *International Journal of Nautical Archaeology* 27 (4): 307–12.

———. 2004. Zhongguo Quanzhou gu-chuan baohu xiufu [Preservation and restoration of the Quanzhou ship in China]. In *The conservation and restoration of Shinan ship: The 20 years history*, ed. National Maritime Museum of Korea, 80–87. Mokpo.

Li, T. 2006. A view from the sea. *Journal of Southeast Asian Studies* 37 (1): 83–102.

Lin, S., G. Du, and J. Green. 1991. Waterfront excavation at Dongmenkou, Ningbo, Zhe Jiang Province, PRC. *International Journal of Nautical Archaeology* 20 (4): 299–311.

Liu, X., H. Li, Y. Chen, X. Yao, and X. Zhang. 2006. Penglai shuicheng xiaohai guchuan caizhi zhuang kuang ji shuzhong peizhi [Quality and use of timbers from Penglai ships]. In *Ancient ships from Penglai*, ed. Cultural Relics and Archaeological Institute of Shandong Province, Yantai Municipal Museum, and Cultural Relics Bureau of Penglai City, 120–38. Beijing: Wenwu Press.

Liu, Z. (ed.). 2011. 2011 *nian Nanhai I hao de kao gu shi jie* [2011 archaeological test excavation of Nanhai No. 1]. Beijing: Science Press.

Lo, J.-P. 1955. The emergence of China as a sea power during the late Sung and early Yuan periods. *Far Eastern Quarterly* 14 (4): 489–503.

Maarleveld, T. J. 1995. Type or technique: Some thoughts on boat and ship finds as indicative of cultural traditions. *International Journal of Nautical Archaeology* 24 (1): 3–7.

Macleod, I. D., and M. Flecker. 2001. Corrosion of tin and its alloys recovered from a 10th century wreck in the Java Sea. In *International Conference on Metals Conservation*, ed. Ian D. Macleod, Johanna M. Theile, and Christian Degrigny, 191–201. Santiago: Western Australian Museum.

Manguin, P.-Y. 1984. Relationship and cross-influence between Southeast Asia and Chinese shipbuilding traditions. In *SPAFA consultative workshop on research on maritime shipping and trade networks in Southeast Asia*, 197–212. Cisarusa: SEAMEO-SPAFA.

———. 1993. Trading ships of the South China Sea: Shipbuilding techniques and their role in the history of the development of Asian trade networks. *Journal of the Economic and Social History of the Orient* 36 (3): 253–80.

———. 2009. The boat remains of Punjulharjo: A preliminary report. Unpublished paper.

———. 2010. New ships for new networks: Trends in shipbuilding in the South China Sea in the 15th and 16th centuries. In *Southeast Asia in the fifteenth century: The China factor*, ed. Geoff Wade and Sun Laichen, 333–58. Singapore: NUS Press.

Manguin, P.-Y., A. Mani, and G. Wade (eds.). 2011. *Early interactions between South and Southeast Asia: Reflections on cross-cultural exchange.* Singapore: Institute of Southeast Asian Studies.

Mathers, W. M., and M. Flecker. 1997. Archaeological report: Archaeological recovery of the Java Sea wreck. Annapolis, Md.: Pacific Sea Resources.

Matsuki, T. 1988. Fune no kigen to hattatsu shou: Kuribune kara kozosen he [Origin and developments of ships: From dugout canoes to built-up ships]. In *Kodai no fune* [Ancient ships], ed. Fukuoka City Museum, 41–42. Fukuoka.

Matsunami, H. 2006. 9 seiki Nippon no koueki katsudou to Chouhoko [Ninth-century maritime activity by Japan and Jang Bogo]. In *A person in the Silla, Jang Bogo*, ed. National Maritime Museum, 143–49. Mokpo.

Matsuura, A. 1995. *Chugoku no kaizoku* [Pirates in China]. Tokyo: Toho Shoten.

———. 2003. *Chinese maritime merchants and pirates*. Libretto of the World History. Tokyo: Yamakawa Shuppansha.

———. 2008a. The activities of Chinese junks on East Asian seas from the seventeenth to the nineteenth centuries: Mainly based on sand junks and bird junks. *Mariner's Mirror* 94 (2): 150–59.

———. 2008b. Modern East Asian cultural interaction and Chinese ships. *Journal of East Asian Cultural Interaction Studies* 1: 41–62.

McCarthy, M. 2005. *Ships' fastenings: From sewn boat to steamship*. College Station: Texas A&M University Press.

McGrail, S. 1985. Classification of water transport. *World Archaeology* 16 (3): 289–303.

———. 2001. *Boats of the world: From the Stone Age to medieval times*. Oxford, U.K.: Oxford University Press.

McKinnon, E. E. 2001. Ancient shipwrecks in Indonesian waters: Lecture to the Oriental Ceramic Society of Indonesia. Jakarta.

Merwin, D. 1977. Selections from Wen-wu on the excavation of a Sung dynasty seagoing vessel in Chuan-chou. *Chinese Sociology and Anthropology* 9 (3): 3–106.

Mikami, T. 1969. *Toji no michi: Tozai bunmei no setten wo tazunete* [The Ceramic Road: Exploring the cultural encounter between East and West]. Tokyo: Iwanami Shoten.

Miksic, J. N. 1983. Srivijaya: Political, economic, and artistic frameworks for analysis. In *SPAFA consultative workshop on archaeological and environmental studies on Srivijaya*, 225–36. Bangkok: SEAMEO-SPAFA.

———. 2010. Before and after Zheng He: Comparing some Southeast Asian archaeological sites of the 14th and 15th centuries. In *Southeast Asia in the fifteenth century: The China factor*, ed. Geoff Wade and Sun Laichen, 384–408. Singapore: NUS Press.

Miksic, J. N., and G. G. Yian (eds.). 2013. *Ancient harbours in Southeast Asia: The archaeology of early harbours and evidence of inter-regional trade*. Bangkok: SEAMEO-SPAFA.

Mills, J. V. 1979. Chinese navigators in Insulinde about A.D. 1500. *Archipel* 18: 69–93.

Ministry of Culture and Fine Arts. 2011. *The preliminary cartography of cultural underwater heritage in Cambodia*. Phnom Penh: Ministry of Culture and Fine Arts and UNESCO Phnom Penh Office.

Mitsutani, T. 1996. Takashima kaitei iseki (Kozaki-ko) shutsudo mokuzai no zyusyu [Wood species of recovered timbers from the Takashima underwater site]. In *Takashima Kaitei Iseki III* [The Takashima underwater site 3], ed. Nagasaki-ken Takashima-cho Board of Education, 73–80. Nagasaki: Nagazaki-ken Takashima-cho Board of Education.

Momoki, S. 1998. Dai Viet and the South China Sea trade from the 10th to the 15th century. *Crossroads: An Interdisciplinary Journal of Southeast Asian Studies* 12 (1): 1–34.

——— (ed.). 2008. *Kaiiki Ajia shi kenkyu nyumon* [A research guide to maritime Asian history]. Tokyo: Iwanami Shoten.

Moon, W.-s. 2011. The introduction of conservation treatment of maritime artifacts in Korea. In *Asia-Pacific Regional Conference on Underwater Cultural Heritage Proceedings*, ed. Mark Staniforth, Jennifer Craig, Sheldon Clyde Jago-on, Bobby Orillaneda, and Ligaya S. P. Lacsina. The MUA Collection. Manila: National Museum of the Philippines. www.themua.org/collections/files/original/63e22f71d8e655f69b515459b0ce444e.pdf.

Mori, T. 2001. Katsuyu chodo shijiko no seisanchi to nendai ni tsuite [Origin and date of the brown glazed vessel with four handles]. In *Takashima kaitei iseki V* [Takashima underwater site 5], ed. Takashima-cho Board of Education, 4–5. Nagasaki: Nagasaki-ken Takashima-cho Board of Education.

———. 2008. Ajia no umi o watatta Ryusen seiji [Longquan celadon wares for maritime trades in Asia]. In *Mono kara mita kaiiki Ajia shi: Mongoru-So Gen jidai no Ajia to Nihon no koryu* [A maritime Asian history from a viewpoint of trading goods: Asian-Japanese exchanges from Mongolia to Song and Yuan dynasties], ed. Yasuhiro Yokkaichi, 97–120. Fukuoka: Kyushu University Press.

Muckelroy, K. 1978. *Maritime Archaeology*. Cambridge, U.K.: Cambridge University Press.

Mukai, K. 2000. Thai Wan Chinsen Sekisai Toujiki no Chosa [Study of ceramics from shipwrecks of the Gulf of Thailand]. *Archaeological Journal of Kanazawa University* 34: 4–5.

Murai, S. 2005. Jisha zoeiryo tosen o minaosu: Boeki, bunka koryu, chinsen [A reconsideration of Jisha-zoeiryo tosen: Trade, cultural exchange and sunken ship]. In *Minatomachi to kaiiki sekai* [Port city and the maritime sphere], ed. Shosuke Murai, 113–43. Minatomachi no Sekaishi [Series world history of the port cities]. Tokyo: Aoki Shoten.

Museum of Overseas Communication History. 1987. Quanzhouwan Song dai haichuan fajue yu yanjiu [Excavation and research of the Song dynasty ship found on Quanzhou Bay]. In *Quanzhouwan Song dai haichuan fajue yu yanjiu* [Excavation and research of the Song dynasty ship found on the Quanzhou Bay], ed. Museum of Overseas Communication History, 3–68. Beijing: Ocean Press.

Nagazumi, Y. 2001. *Shuinsen* [Red seal vessel]. Tokyo: Yoshikawakobunkan.

National Maritime Museum of Korea. 2004. *The conservation and restoration of Shinan ship*. Mokpo.

———. 2005a. *Guide book for permanent exhibits*. Mokpo.

———. 2005b. *The underwater excavation of the sea off Sibidongpado, Gunsan*. Mokpo.

———. 2006a. *Shinan Underwater Relics and 14th Century Asian Marine Trades: International symposium in celebration of the 30th anniversary of the Shinan wreck excavation*. Mokpo.

———. 2006b. *The Shinan wreck I*. Mokpo.
———. 2006c. *The excavation of Anjwa ship, Sinan*. Mokpo.
———. 2008a. *Ansan Daebudo sea site: Submarine excavation*. Report of the Research of Antiquities of the National Maritime Museum 13. Mokpo.
———. 2008b. *Korean ship and shipbuilding tools in the modern times*. Mokpo.
National Research Institute of Maritime Cultural Heritage. 2009. *Taean treasure ship: Submarine excavation*. Report of the Research of Antiquities of the National Research Institute of Maritime Cultural Heritage 17. Mokpo.
———. 2010. *Taean Mado Shipwreck No. 1: Underwater excavation*. Mokpo.
———. 2011a. *Survey report on the waters of Mado*. Mokpo.
———. 2011b. *Taean Mado Shipwreck No. 2: Underwater excavation*. Mokpo.
Needham, J. 1971. *Science and civilisation in China*. Vol. 4, *Physics and physical technology—civil engineering and nautics*. London: Cambridge University Press.
Nishimura, M., and S. Ueda. 2012. Shipbuilding and ships during the Nguyen dynasty (1802–1883). In *Ship culture and phases in East Asian countries: Comparative regionalism and modern Ryukyu*, ed. Hiromichi Okamoto, 63–88. Cultural Interaction Studies of Peripheral-Areas Series 5. Suita: Kansai University.
Noshiro, S., and H. Abe. 2010. Identification of materials of the Quanzhou ship and Samed Nagam ship. In *Shipwreck ASIA: Thematic studies in East Asian maritime archaeology*, ed. Jun Kimura, 74–77. Adelaide: Maritime Archaeology Program.
Oba, O. 1974. Scroll painting of Chinese junks which sailed to Nagasaki in the 18th century and their equipment. *Mariner's Mirror* 60 (4): 351–62.
———. 1980. *Edo jidai no Nitchu hiwa* [Unknown episodes between Japan and China during the Edo period]. Tokyo: Touhou Shoten.
———. 1991. Portraits of Chinese junks painted by Japanese painters in the Edo period. In *International Sailing Ships History Conference*, ed. Shanghai Society of Naval Architecture and Marine Engineering and Marine History Researchers' Association of CSNAME, 5–18. Shanghai.
———. 2001. *Hyochakusen monogatari: Edo jidai no Nitchu koryu* [Story about drifted boats: Exchange between Japan and China during the Edo period]. Tokyo: Iwanami Shoten.
——— (ed.). 2003. *Nagasaki tokanzu shusei* [The compilation of paintings of the Chinese residences (Tokan) in Nagasaki]. Osaka: Kansai Daigaku Shuppanbu.
Ogata, K. 1972. Light woods and heavy woods: Part 2, Heavy woods. *Tropical Forestry* 24: 1–15.
Okamoto, H. 2008a. Several aspects of sea-going vessels of Ryukyu kingdom in Old Ryukyu period. *Journal of East Asian Cultural Interaction Studies* 1: 221–48.
———. 2008b. Structural transformation of Ryukyu kingdom in the 17th and early 18th centuries, as an intersection of cultural interaction. Paper presented at Cultural Reproduction on Its Interface: From the Perspectives of Text, Diplomacy, Otherness, and Tea in East Asia, Osaka.
Okita, M. 2000a. Fune no kokogaku kenkyu shou [Commentary on the archaeological study of ships]. In *Zenkindai Higashi Ajia ni okeru kaiiki koryu seiritsu jyoken ni kansuru kiso teki kenkyu* [Fundamental study about factors that formed maritime

exchange before the premodern period of East Asia], ed. Kaiiki Koryushi Kenkyukai (Society for Historical Research of Maritime Exchange), 21–31. Tokyo: Toyota Foundation.

———. 2000b. Fune no kokogaku myo [About the archaeology of the ship]. In *Zenkindai higashi Ajia ni okeru kaiiki koryu seiritsu zyoken ni kansuru kiso teki kenkyu* [Fundamental study about factors that formed maritime exchange before the premodern period of East Asia], ed. Kaiiki Koryushi Kenkyukai (Society for Historical Research of Maritime Exchange), 21–31. Tokyo: Toyota Foundation.

Okorokov, A. V. 1995. Archaeological finds of ancient dugouts in Russia and the Ukraine. *International Journal of Nautical Archaeology* 24 (1): 33–45.

Omar, M. 1983. A note on the stone wall and earthen causeway at Kota Batu. *Brunei Museum Journal* 5 (3): 27–50.

Ono, R., C. Katagiri, N. Sakagami, H. Kan, H. Miyagi, and Y. Yamamoto. 2013. Yaeyama ni okeru suichuu bunkaisan no genjo to shorai: Ishigaki jima Erabu oki kaitei iseki o chushin ni [The current state and future possibility of underwater cultural heritage in Yaeyama: Underwater sites off Ishigaki Island and in Erabu waters]. *Ishigaki Shiritu Yaeyama Hakubutukan Kiyo* [Bulletin of Ishigaki City Yaeyama Museum] 22: 20–43.

Orillaneda, B. C. 2008. The Santa Cruz, Zambales shipwreck ceramics: Understanding Southeast Asian ceramic trade during the late 15th century C.E. MA thesis, University of the Philippines, Quezon City.

Osawa, M. 2004. Irons in Japan and the Korean Peninsula based on metallurgy. *Bulletin of the National Museum of Japanese History* 110: 89–122.

———. 2005a. Metallurgical analysis on the iron nails from the Takashima underwater site. In *Takashima kaitei iseki XI* [Takashima underwater site 11], ed. Nagasaki-ken Takashima-cho Board of Education, 33–46. Nagasaki.

———. 2005b. One of the forms of iron producing in the Mongol Empire obtained. In *The Avraga site: Preliminary report of the excavation of the palace of Genghis Khan in Mongolia 2001–2004*, ed. Kokugakuin University, 45–62. Niigata: Department of Archaeology, Niigata University.

Pearson, R. 1969. *Archaeology of the Ryukyu Islands: A regional chronology from 3000 B.C. to the historic period*. Honolulu: University of Hawaii Press.

Pearson, R. (ed.). 2009. *Okinawa: The rise of an island kingdom; archaeological and cultural perspectives*. Proceedings of a symposium, Kingdom of the Coral Seas, November 17, 2007, at the School of Oriental and African Studies, University of London. BAR International Series. Oxford, U.K.: Archaeopress.

Pearson, R., M. Li, and G. Li. 2002. Quanzhou archaeology: A brief review. *International Journal of Historical Archaeology* 6 (1): 23–37.

Phann, N. 2011. History and current status of underwater cultural heritage in Cambodia. In *Asia-Pacific Regional Conference on Underwater Cultural Heritage Proceedings*, ed. Mark Staniforth, Jennifer Craig, Sheldon Clyde Jago-on, Bobby Orillaneda, and Ligaya S. P. Lacsina. The MUA Collection. Manila: National Museum of the Philippines. www.themua.org/collections/files/original/d77fe4436f35606533d74d9fe994f70a.pdf.

Prishanchit, S. 1992. Current movement of underwater archaeology in Thailand. In *Final report of the SEAMEO-SPAFA consultative workshop on underwater archaeological research*, 50–56. Jakarta: SEAMEO-SPAFA.

Priyanto, W. A. 2011. Conservation research and treatment programs: Case study of ancient boat site in Rembang Regency. In *Asia-Pacific Regional Conference on Underwater Cultural Heritage Proceedings*, ed. Mark Staniforth, Jennifer Craig, Sheldon Clyde Jago-on, Bobby Orillaneda, and Ligaya S. P. Lacsina. The MUA Collection. Manila: National Museum of the Philippines. www.themua.org/collections/files/original/44e39965184f60a140f8f92e40bb1500.pdf.

Promboon, S. 1984. Country report of Thailand. In *SPAFA consultative workshop on research on maritime shipping and trade networks in Southeast Asia*, 107–31. Cisarusa: SEAMEO-SPAFA.

Ptak, R. 1998. Quanzhou to the Sulu Zone and beyond: Questions related to the early fourteenth century. *Journal of Southeast Asian Studies* 29 (2): 269–94.

———. 2000. Camphor in East and Southeast Asian trade, c. 1500: A synthesis of Portuguese and Asian sources. In *Vasco da Gama and the linking of Europe and Asia*, ed. Anthony Disney and Emily Booth, 142–66. New Delhi: Oxford University Press.

Reid, A. 1988. *Southeast Asia in the age of commerce, 1450–1680*. Vol. 1, *The lands below the winds*. New Haven, Conn.: Yale University Press.

———. 1995. *Southeast Asia in the age of commerce, 1450–1680*. Vol. 2, *Expansion and crisis*. New Haven, Conn.: Yale University Press.

———. 2010. Hybrid identities in the 15th-century straits. In *Southeast Asia in the fifteenth century: The China factor*, ed. Geoff Wade and Sun Laichen, 307–32. Singapore: NUS Press.

Reinders, R., and K. Paul (eds.). 1991. *Carvel construction technique: Fifth International Symposium on Boat and Ship Archaeology, Amsterdam, 1988*. Oxford, U.K.: Oxbow Books.

Reischauer, E. O. 1940. Notes on T'ang dynasty sea routes. *Harvard Journal of Asiatic Studies* 5 (2): 142–64.

———. 1955. *Ennin's travels in T'ang China*. New York: Ronald Press.

Ridho, A., and E. E. McKinnon. 1998. *The Pulau Buaya wreck*. The Ceramic Society of Indonesia Monograph Series. Jakarta.

Ronquillo, W. P. 1987. The Butuan archaeological finds: Profound implications for Philippines and Southeast Asian prehistory. *Man and Culture in Oceania* 3: 71–78.

Rossabi, M. 1988. *Khubilai Khan: His life and times*. Berkeley: University of California Press.

Sandström, M., F. Jalilehvand, E. Damian, Y. Fors, U. Gelius, M. Jones, and M. Salomé. 2005. Sulfur accumulation in the timbers of King Henry VIII's warship *Mary Rose*: A pathway in the sulfur cycle of conservation concern. *PNAS* 102 (40): 14165–70.

Sandström, M., F. Jalilehvand, I. Persson, U. Gelius, P. Frank, and I Hall-Roth. 2002. Deterioration of the seventeenth-century warship *Vasa* by internal formation of sulphuric acid. *Nature* 415: 893–97.

Sasada, T. 2009. Hokuto Asia no tetsu seisan [Iron production in northeastern Asia]. In *Chusei Higashi Ajia no shuhen sekai* [Peripheries of East Asia in the medieval age], ed. Tetsuya Amano, Eishi Ikeda, and Isao Usuki, 74–85. Tokyo: Douseisha.

Sasaki, R. 2008. The origin of the lost fleet of the Mongol Empire. MA thesis, Texas A&M University, College Station.

———. 2015. *The origins of the lost fleet of the Mongol Empire.* College Station: Texas A&M University Press.

Sasaki, R., and C. H. Lee. 2010. Goryeo dynasty (918–1392) shipwrecks in Korea. In *Shipwreck ASIA: Thematic studies in East Asian maritime archaeology*, ed. Jun Kimura, 56–73. Adelaide: Maritime Archaeology Program.

Schottenhammer, A. 1999. Local political-economic particulars of the Quanzhou region during the tenth century. *Journal of Song Yuan Studies* 29: 1–41.

———. 2010. Transfer of Xiangyao 香藥 from Iran and Arabia to China: A reinvestigation of entries in the Youyang zazu 酉陽雜俎 (863). In *Aspects of the maritime silk road: From the Persian Gulf to the East China Sea*, ed. Ralph Kauz, 117–49. Wiesbaden: Harrassowitz Verlag.

Shan, J. 2011. From underwater archaeology to underwater cultural heritage protection: Speech for the International Meeting on the Protection, Presentation and Valorisation of Underwater Cultural Heritage. In *2010 International Meeting on Protection, Presentation and Valorisation of Underwater Cultural Heritage*, ed. Shuguang Liu, 199–205. Beijing: Wenwu Press.

Shanxi Museum and Hainan Provincial Museum. 2013. *Huaguang jiao I hao shenchuan yizhen* [Huaguang reef shipwreck no. 1 site]. Shanxi: Shanxi Renmin Chubanshe.

Shiba, Y. 1968. *Sodai shoogyooshi kenkyu: Commercial activities during the Sung dynasty.* Tokyo: Kazamashobo.

Shiga Prefectural Association for Cultural Heritage (ed.). 2006. *Marukibune no jidai: Biwako to kodaijin* [Time period of the dugout canoe: Lake Biwa and ancient people]. Otsu: Shiga Prefectural Association for Cultural Heritage.

Shimizu, J. 1968. Kodai no fune: Nihon no marukibune o chushin ni [Ancient ships: From the study of dugout canoes in Japan]. In *Fune* [Ship], ed. Toshiichi Sudo, 31–55. Tokyo: Hosei University Press.

———. 1975. Nihon kodai no fune [Ancient Japanese ships]. In *Fune: Nihon kodai bunka no tankyu* [Ship: Study on Japanese ancient culture], ed. Taryo Obayashi, 12–83. Tokyo: Shakai Shisosha.

Sigaut, E. 1960. A northern type of Chinese junk. *Mariner's Mirror* 46 (3): 161–74.

Sleeswyk, A. W. 1996. The Liao and the displacement of ships in the Ming navy. *Mariner's Mirror* 82 (1): 3–13.

Sokoloff, V. A. 1982. *Navios da China* [Ships of China]. Macao: Banco Nacional Ultramarino.

Staniforth, M. 2002. *Material culture and consumer society: Dependent colonies in colonial Australia.* The Plenum Series in Underwater Archaeology. New York: Kluwer Academic/Plenum Publishers.

Stargardt, J. 1984. The isthmus of the Malay Peninsula in long-distance navigation: New archaeological findings. In *SPAFA consultative workshop on research on maritime shipping and trade networks in Southeast Asia*, 1–25. Cisarua: SEAMEO-SPAFA.

Steffy, J. R. 1994. *Wooden ship building and the interpretation of shipwrecks.* College Station: Texas A&M University Press.

Sudaryadi, A. 2011. The Belitung shipwreck site after commercial salvage in 1998. In *Asia-Pacific Regional Conference on Underwater Cultural Heritage Proceedings*, ed. Mark Staniforth, Jennifer Craig, Sheldon Clyde Jago-on, Bobby Orillaneda, and Ligaya S. P. Lacsina. The MUA Collection. Manila: National Museum of the Philippines. www.themua.org/collections/files/original/96327166d24100d76d20450925da1524.pdf.

Suzuki, K., Y. Karakida, and Y. Kamada. 2000. Provenance of granitic anchor stones recovered from the Takashima submerged site: An approach using the CHIME method for dating of zircons. *Proceedings of the Japan Academy Series B, Physical and Biological Sciences* 76 (9): 139–44.

Takashima-cho Board of Education. 1996. *Takashima kaitei iseki III* [Takashima underwater site 3]. Takashima-cho bunkazai chosa hokokusho [A series survey report of cultural properties in Takashima-cho]. Nagasaki: Nagasaki-ken Takashima-cho Board of Education.

———. 2001. *Takashima kaitei iseki V* [Takashima underwater site 5]. Takashima-cho bunkazai chosa hokokusho [A series survey report of cultural properties in Takashima-cho]. Nagasaki: Nagasaki-ken Takashima-cho Board of Education.

———. 2003. *Takashima kaitei iseki VIII* [Takashima underwater site 8]. Takashima-cho bunkazai chosa hokokusho [A series survey report of cultural properties in Takashima-cho]. Nagasaki: Nagasaki-ken Takashima-cho Board of Education.

———. 2008. *Matsuura-shi Takashima kaitei iseki* [Matsuura City, Takashima underwater site]. Matsuura-shi bunkazai chosa hokokusho [A series survey report of cultural properties in Matsuura City]. Matsuura: Nagazaki-ken Matsuura-shi Takashima-cho Board of Education.

Tanaka, K., and A. Matsuura. 1986. *Materials concerning the Chinese ship cast away on the coast of Sizuoka Prefecture in 1826 (Bunsei 9)*. Materials Series of Chinese Ships Cast Away on the Coast of Japanese Islands. Osaka: Kansai Daigaku Shuppanbu.

Tian, X., N. Li, and Z. Zhang. 2014. Study of the corrosion of copper coins from the wreck of "The South China Sea No. 1." In *Proceedings of the 2nd Asia-Pacific Regional Conference on Underwater Cultural Heritage*, ed. Hans Van Tilburg, Sila Tripati, Veronica Walker Vadillo, Brian Fahy, and Jun Kimura. The MUA Collection. Honolulu: Museum of Underwater Archaeology. www.themua.org/collections/files/original/cac6784e358b17b4e9cc00d8dddc5e94.pdf.

Tomiyama, K. 2012. Ships and the history of the Ryukyus: Various aspects of Ryukyuan ships in the early modern period. In *Ship culture and phases in East Asian countries: Comparative regionalism and modern Ryukyu*, ed. Hiromichi Okamoto, 23–35. Suita: Kansai University.

Tripati, S., A. S. Gaur, and Sundaresh. 2004. Marine archaeology in India. *Man and Environment* 29 (1): 28–41.

Tripati, S., A. S. Gaur, Sundaresh, and S. N. Bandodker. 2001. Exploration for shipwrecks off Sunchi Reef, Goa, west coast of India. *World Archaeology* 32 (3): 355–67.

Twitchett, D., and J. Stargardt. 2002. Chinese silver bullion in a tenth-century Indonesian wreck. *Asia Major* 15 (1): 23–72.

Uken-son Board of Education. 1999. *Kagoshima-ken Oshima-gun Uken-son Kurakizaki kaitei iseki hakkutsu chosa hokokusho* [Survey and excavation report at the Kurakizaki underwater site off the coast of Uken village in Oshima district, Kagoshima prefecture]. Uken-son bunkazai chosa hokokusho No.2 [A series survey report of cultural properties in Uken-son No.2]. Kagoshima: Uken-son Board of Education.

Underwood, H. H. 1979. Korean boats and ships. In *English Publication no. 2, The Literary Department, Chosen Christian College*. Seoul: Yonsei University Press.

Van Tilburg, H. 1994. Chinese maritime history and nautical archaeology: Where have all the ships gone? *Bulletin of the Australasian Institute for Maritime Archaeology* 18 (2): 7–10.

———. 2007. *Chinese junks on the Pacific: Views from a different deck*. New Perspectives on Maritime History and Nautical Archaeology. Gainesville: University Press of Florida.

Vatcharangkul, E. 1992. Treasure-hunter, volunteer and researcher: Defining assets for underwater heritage research. In *Final report of the SEAMEO-SPAFA consultative workshop on underwater archaeological research*, 57–64. Jakarta: SEAMEO-SPAFA.

Viraphol, S. 1977. *Tribute and profit: Sino-Siamese trade, 1652–1853*. Harvard East Asian Monographs. Cambridge, Mass.: Harvard University Press.

Wade, G. 2003. *The pre-modern East Asian maritime realm: An overview of European-language studies*. Asia Research Institute Working Paper Series. Singapore: Asia Research Institute, National University of Singapore.

———. 2004. *The Zheng He voyages: A reassessment*. Asia Research Institute Working Paper Series. Singapore: Asia Research Institute, National University of Singapore.

———. 2007. *Ryukyu in the Ming reign annals, 1380s–1580s*. Asia Research Institute Working Paper Series. Singapore: Asia Research Institute, National University of Singapore.

———. 2010. Southeast Asia in the 15th century. In *Southeast Asia in the fifteenth century: The China factor*, ed. Geoff Wade and Sun Laichen, 3–43. Singapore: NUS Press.

Wagner, D. B. 1995. The traditional Chinese iron industry and its modern fate. *Chinese Science* 12: 138–61.

———. 2001. Blast furnaces in Song-Yuan China. *East Asian Science, Technology, and Medicine* 18: 41–74.

Wang, Guanzhou. 2000. *Zhongguo gu chuan tu pu* [Catalogue of ancient Chinese ships]. Beijing: Sanlian Shudian.

Wang, Gungwu. 1958. The Nanhai Trade: A study of the early history of Chinese trade in the South China Sea. *Journal of the Malayan Branch Royal Asiatic Society* 31 (2): 1–135.

Watanabe, T. 1985. Ancient Cambodia Funan as described by Zhu Ying and Kang Tai: A collection of fragments written by a mission to Cambodia in Three Kingdoms Wu period. *Proceedings of the Faculty of Letters of Tokai University* 43: 7–28.

Waters, D. W. 1938. Chinese junks: Antung trader. *Mariner's Mirror* 24: 49–67.

———. 1939. Chinese junks: The Pechili trader. *Mariner's Mirror* 25: 62–88.

———. 1946. Chinese junks: The *twaqo*. *Mariner's Mirror* 32: 155–67.

———. 1947. Chinese junks: The Hangchow Bay trader and fisher. *Mariner's Mirror* 33: 28–38.

Wei, J. 2011. Innovative thoughts on the preservation of underwater cultural heritage in China: No. 1 Nanhai as a project example. In *2010 International Meeting on Protection,*

Presentation and Valorisation of Underwater Cultural Heritage, ed. Shuguang Liu, 357–71. Beijing: Wenwu Press.

Worcester, G.R.G. 1941. *Notes on the crooked-bow and crooked-stern junks of Szechwan*. Maritime Customs, Miscellaneous Series 3. Shanghai: Statistical Department of the Inspectorate General of Customs.

———. 1966. *Sail and sweep in China: The history and development of the Chinese junk as illustrated by the collection of junk models in the science museum*. London: Science Museum.

———. 1971. *The junks and sampans of the Yangtze*. Annapolis, Md.: United States Naval Institute.

Wu, C. 2011. Archaeological and ethnological research pertaining to underwater cultural heritage in China's surrounding seas. In *2010 International Meeting on Protection, Presentation and Valorisation of Underwater Cultural Heritage*, ed. Shuguang Liu, 245–63. Beijing: Wenwu Press.

Xi, L. (ed.). 1989. *Penglai gunchuan yu Dengzhou gugang* [The ancient Penglai ship and the ancient Dengzhou port]. Dalian: Dalian Haiyun Xueyuan Chubanshe.

———. 2000. *Zhongguo zaochuan shi* [History of Chinese shipbuilding]. Hubei: Hubei Education Publisher.

——— (ed.). 2008. *Chuan wenhua* [Ship culture]. Beijing: China Communication Press.

Xi, L., and Y. Xin. 1991. Preliminary research on the historical period and restoration design of the ancient ship unearthed in Penglai. In *International Sailing Ships History Conference*, ed. Shanghai Society of Naval Architecture and Marine Engineering and Marine History Researchers' Association of CSNAME, 225–36. Shanghai.

Xi, L., X. Yang, and X. Tang (eds.). 2004. *Zhongguo kexue jishu shi: Jiaotong juan* [The history of science and technology in China: Transportation]. Beijing: Science Press.

Xu, Y. 1985. Origin and technique of "gua-ju" (iron cramp) connections in wooden craft construction. *Marine History Research, Transactions of CSNAME Marine History Research Group* 1: 64–71.

Yagi, H., H. Kawabe, Y. Terao, K. Wanatabe, and H. Adachi. 2008. The hydrodynamic performance of a Ryukyuan tribute ship. In *The East Asian "Mediterranean": Maritime crossroads of culture, commerce and human migration*, ed. Angela Schottenhammer, 235–51. Wiesbaden: Harrassowitz Verlag.

Yahya, M.B.H. 2005. *The boats of Brunei Darussalam*. Bandar Seri Begawan: Brunei Museums Department.

Yamamoto, T. 1980. Chinese activities in the Indian Ocean before the coming of the Portuguese. *Diogenes* 3: 19–34.

———. 1983. Reexamination of historical texts concerning Srivijaya. In *SPAFA consultative workshop on archaeological and environmental studies on Srivijaya*, 201–209. Bangkok: SEAMEO-SPAFA.

Yamauchi, S. 2008. Nihon Reto to kaiiki sekai [Japan archipelago and maritime space]. In *Kaiiki Ajia shi kenkyu nyumon* [A research guide to maritime Asian history], ed. Shiro Momoki, 39–47. Tokyo: Iwanami Shoten.

Yang, S.-S. 2011. Vessel cargo loading methods of the Goryeo period. In *Asia-Pacific Regional Conference on Underwater Cultural Heritage Proceedings*, ed. Mark Staniforth,

Jennifer Craig, Sheldon Clyde Jago-on, Bobby Orillaneda, and Ligaya S. P. Lacsina. The MUA Collection. Manila: National Museum of the Philippines. www.themua.org/collections/files/original/ed9140b6aced83755115891935249f33.pdf.

Yang, X. 2009. China's junk construction technology. *Maritime History Studies*, no. 1: 55–70.

Yang, Y. 1989. Maritime fortress Penglai in Shandong Province and Ming combat vessel. In *Penglai guchuan yu dengzhou gugang* [The ancient Penglai ship and the ancient Dengzhou port], ed. Longfei Xi, 60–66. Dalian: Dalian Haiyun Xueyuan Chubanshe.

Yang, Z. 2009. An introduction to the hull of Huaguang Reef One shipwreck site of the Song dynasty in Xisha Islands. Paper presented at Of Ships and Men, the Sino-French Workshop, Ningbo.

Yasuhiro, Y. 2008. Chinese and Muslim diasporas and the Indian Ocean trade network under Mongol hegemony. In *The East Asian "Mediterranean": Maritime crossroads of culture, commerce and human migration*, ed. Angela Schottenhammer, 73–102. East Asian Maritime History. Wiesbaden: Harrassowitz Verlag.

Yiran, F. 2014. Underwater cultural heritage conservation and the convention practice in China. In *The Proceedings of the 2nd Asia-Pacific Regional Conference on Underwater Cultural Heritage*, ed. Hans Van Tilburg, Sila Tripati, Veronica Walker Vadillo, Brian Fahy, and Jun Kimura. The MUA Collection. Honolulu: Museum of Underwater Archaeology. www.themua.org/collections/files/original/748c7069c5d4151dcddadfb0918b8709.pdf.

Young, C.F.T. 1867. *The fouling and corrosion of iron ships: Their causes and means of prevention, with the mode of application to the existing iron-clads*. London: London Drawing Association.

Yuan, X. 2006a. The excavation and accomplishment of the shipwreck on Penglai area in China. In *Shinan Underwater Relics and 14th Century Asian Marine Trades: International symposium in celebration of the 30th anniversary of the Shinan wreck excavation*, ed. National Maritime Museum of Korea, 471–513. Mokpo.

———. 2006b. Three ancient boats unearthed from Penglai, Shandong: A brief report. *Maritime History Studies*, no. 1: 1–19.

Yule, C. H. 1882. Notes on the oldest records of the sea-route to China from western Asia. *Proceedings of the Royal Geographical Society and Monthly Record of Geography* 4 (11): 649–60.

Yule, H. 1993. *The travels of Marco Polo: The complete Yule-Cordier edition*. New York: Courier Dover Publications.

Zhang, C., Z. You, Z. Wu, and Y. Liu (eds.). 2004. *Zhongguo jixie gongchen faming shi* [History of technological and mechanical invention in China]. Beijing: Singh University Press.

Zhang, W. (ed.). 2001. *Suizhong Sandaogang Yuandai Chenchuan* [Yuan dynasty shipwreck in Suizhong Shandaogang]. Beijing: Science Press.

———. 2006a. The application of the basic principles and methods of the underwater archaeology in China. In *Shinan Underwater Relics and 14th Century Asian Marine Trades: International symposium in celebration of the 30th anniversary of the Shinan wreck excavation*, ed. National Maritime Museum of Korea, 429–34. Mokpo.

———. 2006b. Preface. In *Xisha shuixia kaogu (1998–1999)* [Underwater archaeology in Xisha Islands], ed. Underwater Archaeology Research Center of the National Museum of China and Cultural Heritage Administration Office of Hainan Province, v–xi. Series of Underwater Archaeology Report. Beijing: Science Press.

———. 2011. The origins and creation of China's underwater archaeology. In *2010 International Meeting on Protection, Presentation and Valorisation of Underwater Cultural Heritage*, ed. Shuguang Liu, 223–34. Beijing: Wenwu Press.

Zhao, H. 1992. Recent development in the legal protection of historic shipwrecks in China. *Ocean Development and International Law* 23: 305–33.

Zheng, Y. 1991. Zheng He's naval expeditions and the development of marine economy of early Ming dynasty. In *International Sailing Ships History Conference*, ed. Shanghai Society of Naval Architecture and Marine Engineering and Marine History Researchers' Association of CSNAME, 191–96. Shanghai.

Zou, Y., and X. Yuan. 1989. A study of date and use of ancient Penglai ship. In *Penglai guchuan yu Dengzhou gugang* [The ancient Penglai ship and the ancient Dengzhou port], ed. Longfei Xi, 74–79, 177. Dalian: Dalian Haiyun Xueyuan Chubanshe.

Index

Page numbers in *italics* refer to illustrations and tables.

Jun Kimura is junior associate professor in the Department of Maritime Civilizations at Tokai University and an advisory member of the Japanese government's Research and Examination Committee of Underwater Heritage. He is also an advisory editorial member of the *International Journal of Nautical Archaeology*. He has served as a principal investigator on the Kublai Khan naval battlefield site in Vietnam.

www.ingramcontent.com/pod-product-compliance
Lightning Source LLC
Chambersburg PA
CBHW031932100926
51263CB00002B/91
9780813061184